D0914612

BIRD MIGRATION

BY

DONALD R. GRIFFIN

Illustrations by Helen C. Lyman

DOVER PUBLICATIONS, INC.

NEW YORK

Copyright © 1964 by Donald R. Griffin.
Copyright © 1974 by Dover Publications, Inc.
All rights reserved under Pan American and
International Copyright Conventions.

Published in Canada by General Publishing Company, Ltd., 30 Lesmill Road, Don Mills, Toronto,
Ontario.
Published in the United Kingdom by Constable
and Company, Ltd., 10 Orange Street, London WC 2.

This Dover edition, first published in 1974, is an
unabridged, slightly corrected republication of the
work originally published by Doubleday & Company,
Inc., Garden City, New York, in 1964 as part of the
Anchor Books "Science Study Series." The author
has prepared a new Preface and updated Further
Reading especially for the Dover edition.

International Standard Book Number: 0-486-20529-0
Library of Congress Catalog Card Number: 74-76321

Manufactured in the United States of America
Dover Publications, Inc.
180 Varick Street
New York, N. Y. 10014

PREFACE TO THE DOVER EDITION

Biologists have learned a great deal about bird migration since this book was first published in 1964. While orientation based on the sun and stars is unquestionably important to birds, improved radar observations have shown that birds possess reasonably good orientation when flying under, inside or between thick layers of cloud. While most birds prefer to fly on clear nights, and while their flight directions are more variable when they cannot see the stars, they must have a means of navigation that does not depend on a clear view of either sky or ground.

Exciting new and ingenious experiments with homing pigeons by William Keeton of Cornell University and his colleagues have demonstrated orientation based on something other than the sun. Contrary to the results of most previous experiments, Keeton found that pigeons that had been appropriately trained and motivated had an accurate initial orientation toward home under overcast skies. Furthermore, when the biological clocks of these pigeons were shifted by approximately six hours, the effect on the birds' initial headings was very different from that shown in Matthews' experiments with mallard ducks (described in Chap-

ter 8). If the clock-shifted pigeons could see the
sun, their initial headings were deflected as shown
in the upper portion of Figure 27 (p. 164). But
other pigeons, subjected to the same clock-shifting
treatment and released under overcast skies,
headed almost as accurately toward home as un-
treated control birds.

New experiments have convinced many biologists that birds use the earth's magnetic field for
orientation. The effects of altering the magnetic
field around the birds have so far been slight and
variable. The entire subject—too complex for brief
discussion—has been well reviewed by Keeton and
Emlen, whose ideas deserve the most serious consideration.

Radar has also helped to explain the previously
puzzling phenomenon known as "reversed migration," in which birds move in directions that are
not what we would consider appropriate for the
season. Radar observations show that the smaller
species of birds, whose speed of flight is comparable to wind speeds, almost always migrate downwind. Flight in the same direction as the wind
often causes birds to move in the "wrong" direction—south in the spring and north in the fall.

Finally, the concept of genetically influenced
orientation behavior has been refined through important experiments by Emlen. Rather than a detailed star map being coded in the genes, it now
seems more likely that the "genetic instructions"
take approximately the following form for birds
in the Northern Hemisphere: "Watch the motions
of the stars and notice which part of the sky
changes relatively little during the night. Then,
in the autumn, fly away from this unchanging
group of stars."

Recent advances have enabled us to understand bird migration a little better, and knowing more about birds makes them even more fascinating and impressive. Further Readings (pp. 174–175) has been revised to reflect these advances.

D.R.G.

New York City
February, 1974

PREFACE TO THE ORIGINAL EDITION

In late summer it is common to see hundreds of barn swallows perched in rows on telephone or electric power lines. They show no concern whether kilowatts of electric power are flowing between their toes—or farmhouse gossip. Both types of wires serve as equally convenient substitutes for the tree limbs that have supported barn swallows at this season for millions of years. A scientist applying his methods to the scene can treat the wires as uniform, attenuated metal cylinders conducting electricity at a clearly defined rate, and dismiss the poles, the roadway, and the swallows as quite irrelevant to the physics of power transmission or the capacity of the wires to carry information over a given distance. But as he goes about setting his thoughts in order, he finds the birds themselves less easily disposed of. Their mass and temperature and the rate at which they release energy by oxidizing foods within their bodies have all been measured, if not in barn swallows, in so many other animals that these properties of a barn swallow as it rests upon the wires can be estimated with considerable confidence. But just as the contemplative scientist begins to sort out such thoughts, the birds take off in a twittering crescendo, circle in a close flock over the neighboring fields, and then assemble again on other wires some distance down the roadway.

Why these short excited flights by the whole flock of swallows? And why were there no such aggregations of barn swallows on wires earlier in the summer? Perhaps the scientist is called to dinner before he can make any further sense out of these facts, but when he returns the following evening he finds the wires quite empty. He may drive along miles of country roads scanning identical wires to no avail; the barn swallows have vanished. A little inquiry discloses that these are highly migratory birds; they feed on insects, and, of course, cold weather removes this source of food. But long before the insects diminish in numbers, the swallows will have flown thousands of miles south to tropical regions where insects are abundant throughout the northern winter.

How can a feathered flying machine that weighs fifteen grams (about half an ounce) migrate three to five thousand miles twice each year, catching and digesting its fuel supply along the way? Why do the swallows fly south before their food supply is in any way diminished? And why, having spent the winter in a climate where insect food is available throughout the year, do they fly north again in spring? What sort of flying machine *is* a bird, after all, and how do its operating machinery and its efficiency compare with aircraft designed and built by men? A migrating bird, even when it is considered merely as a physical system, differs from any artificial machine in the multitude of complex mechanisms that are squeezed into a very small space. The aerodynamics of bird flight is much more intricate than that of airplanes and more impressive in almost every respect except for speed. The whole shape of a bird's wing changes throughout the cycle of its wingbeats, and yet at each stage every feather is working effectively to achieve lift or propulsion. Turns, take-offs, and landings require still more refined adjustments, and these must be accomplished speedily since small birds beat their wings many times per second.

Migrating birds also present us with baffling problems of navigation. How do they extract information about a distant goal from the sky or landscape, which to a human observer seems to have no relation whatever to the rain forests of the Amazon Valley or to the lake-studded tundras of Alaska? Most of the smaller birds migrate at night, often crossing hundreds of miles of open ocean. Yet to the migrating bird something speaks out from the earth or sky with directions to guide its journey.

Bird migrations illustrate the challenges of biophysics. How far have physical principles carried us toward an understanding of the complex problems of biology? Not all the answers are available, but in the following chapters we can review what has been accomplished, and what those scientists who have considered the matter most carefully believe are the most likely solutions. One of the most tantalizing aspects of the subject is that new ideas are called for, and fresh minds approaching old questions are as likely as any to see the way to new advances.

CONTENTS

Chapter 1

THE EXTENT OF BIRD MIGRATIONS

Birds appear to differ so widely in their powers of flight that really long migrations over thousands of miles, from one continent to another, would seem possible only for such sizable and strong fliers as ducks, geese, gulls, or hawks. But it turns out on closer study that certain hummingbirds weighing only three or four grams (about one eighth ounce) migrate farther than swans and pelicans, 2500 times as large. Indeed it is exceptional, at least in temperate climates, for any wild bird to spend its entire life within a few miles of the nest where it hatched from an egg. Bird migrations are impressive in magnitude, distances covered, numbers of individual birds involved, and altitudes of flight. Before turning our attention to other fascinating aspects of migration, it is helpful to survey the sheer extent of these journeys in time and space.

The Local Appearance and Disappearance of Species

How have biologists learned such facts about bird migrations as the distances covered by a given species or individual? The oldest method is still the basis for most of our knowledge of bird migrations, despite its simplicity. It is to observe which species are present in various localities throughout the year, when

they appear and disappear, and how their numbers
fluctuate. Most birds do not display themselves as
conspicuously as swallows on telephone wires; they
must be sought out in swamps or thickets or where-
ever they may choose to spend their time. Many
species that look quite similar have altogether different migratory habits, and this is one reason why bird
watchers pay such close attention to identifying the
many species that fill out their checklists. The longest and most interesting migration routes have been
worked out by the tedious compiling and comparing
of literally thousands of observations that a particular species was or was not present at a given place
on a certain date. For example, the barn swallows of
the September telephone wires disappear on approximately the following dates from a few typical states
and Canadian provinces where they have been
closely studied: Saskatchewan, September 22; North
Dakota, September 28; Missouri, October 11; and
Louisiana, November 3. In winter barn swallows have
been seen only rarely north of the mainland of South
America. Most of them spend the winter months in
Columbia, Brazil, Peru, Bolivia, Paraguay, and northern Argentina. On the way north in spring they are
usually seen in Louisiana about March 20; Missouri,
April 7; North Dakota, April 25; and Saskatchewan,
April 30. Of course, there is variation from year to
year in the exact date when the first swallow is observed, and also between the warmer and colder parts
of the states mentioned. Nor do all arrive at once;
their numbers rise and fall gradually, although there
are often sudden arrivals and departures of large
flocks.

The summer and winter ranges of the barn swallow
are shown in Fig. 1, and there is no doubt that an
extensive migration is required twice each year to

Fig. 1. Summer and winter ranges of the barn swallow.

take this population of birds from North to South America and back again. But one cannot tell from this sort of seasonal distribution whether the swallows from Saskatchewan winter in Columbia or in Paraguay. And for many species of birds the summer and winter ranges are not so clearly separated, while for others they overlap extensively. A good example of the latter is the common crow, found in summer all over the United States and most of Canada and Alaska north to the limit of wooded regions. In winter crows are absent from all but the most southern parts of Canada, but are still to be found in much of the United States. Obviously those that spent the summer, and raised their young, in northern Canada must have migrated south; but observation of seasonal distribution provides only inconclusive information about areas where they are present throughout the year. Although overlapping summer and winter ranges are very common among birds, much patient observation and compilation of records have demonstrated that migrations are underway, even though the particular species is present at all seasons. These signs of migration may be temporary disappearances for some weeks, followed by a new period of abundance. Or the numbers of a particular kind of bird may fluctuate suddenly. Finally, the birds can often be seen flying northward in spring or south in fall, singly, or in flocks. Each of these observations taken by itself can be interpreted in other ways, but the whole picture establishes the fact that birds are migrating.

The Banding of Birds

This general picture of bird migration, based on dated observations of presence or absence, has been substantiated by more direct and specific methods. It is a convenient characteristic of birds' legs that they are covered with hard scales, and that loosely fitting cylindrical bands can be attached without causing irritation or inconvenience. The enlargement at the toes prevents the metal or plastic band from slipping off. Sometimes colored bands are used to distinguish a small number of groups of birds, so that their travels can be traced without the need for catching each one individually. Some species can be marked as fledglings while still in their nests, but most bird banding is done by trapping or netting adults or well-grown young, and releasing them unharmed within a few minutes. Roughly fifteen million birds have been banded over the past sixty years in North America and Europe, and each band carries not only an identifying number but a return address to which information should be sent if the bird is found elsewhere. In North America most bands are inscribed, NOTIFY FISH AND WILDLIFE SERVICE, WASHINGTON, D. C., or some abbreviation of this address.

The proportion of banded birds recovered at any significant distance varies widely with the species. Ducks and geese that are hunted heavily yield recoveries up to 20 or 25 per cent, but only a fraction of 1 per cent of the banded, small songbirds are ever recovered at distant points, and even fewer of those species that spend most of their time flying over the ocean. When a banded bird is recovered, this fact establishes only two points on what may have been a migratory journey of much greater distance than the straight line drawn between them on a map. Usually some time elapses between banding and recovery at any great distance, often an interval of several years. This tells something about the bird's life span, but very little about the details of any one migration. For example, a barn swallow banded in Saskatchewan was retaken in Bolivia—impressive evidence of the extent of its migration. But it was not recovered until six years after being banded, and in the interval it had made eleven migrations between North and South America.

In spite of these limitations, a few out of the

thousands of recoveries of banded birds have involved sufficiently fortunate circumstances and timing to yield a fairly accurate picture of specific migratory flights. For example, a barn swallow banded in Massachusetts on June 28 was found dead (trapped in the asphalt of a tar roof) in Florida on August 26 of the same year. We may regret this bird's misfortune, but nevertheless take advantage of the information made available by the recovery of its band. Since barn swallows do not leave Massachusetts until late summer, the time actually taken for the 1225-mile flight was doubtless far less than the two months that elapsed between banding and recovery. On the other hand, a few recovery records have demonstrated some approximation to the actual speed and some substantial fraction of the total migration. For example, a certain purple finch was banded in central Massachusetts on its northward migration in spring, and was retaken three days later in Bar Harbor, Maine. Purple finches are seed-eating birds related to sparrows, and their migrations often are little longer than the 230 miles separating these points of banding and recovery. It need not diminish our respect for this bird to note that since its flight speed is about twenty to twenty-five miles per hour, the trip could have been accomplished in a few hours of flight each day.

Some Spectacular Migrations

One of the smaller species of duck, the blue-winged teal, has turned in some of the most rapid long-distance flight records available. In less than twenty-seven days one of these birds flew from the St. Lawrence River, near Quebec, to British Guiana. The distance between points of banding and recovery is 3300 miles; hence, the minimum average rate of travel was 122 miles per day. Even more rapid flights by this species under slightly unusual conditions are described in Chapter 8. The shore birds or waders, comprising various sandpipers, plovers, and their relatives, are all rapid fliers, and they carry out some of the most impressive migrations. A medium-sized sandpiper belonging to a species called the dowitcher traveled from the coast of Massachusetts to the Panama Canal Zone between August 24 and September 12, covering this 2300-mile stretch at an average minimum rate of 125 miles per day. A semi-palmated sandpiper, weighing about 15 grams, or half an ounce, flew 2400 miles from Massachusetts to Venezuela in twenty-six days, or an average of 92 miles per day. But the most rapid shore-bird flight documented by banding was that of a lesser yellow-legs weighing roughly 100 grams, or about 3.5 ounces. It was trapped and banded on the coast of Massachusetts on August 28, and shot on the island of Martinique, in the West Indies, the following September 3. This bird covered 1930 miles in six days. Even if it flew a perfectly straight course, and departed immediately after its release, only to be shot the moment it arrived on the shore of Martinique, this bird averaged 322 miles per day.

These records were obtained from large-scale banding of certain species that happen to congregate in large numbers where biologists could make a sustained effort to trap and band them. There are other species of birds, less readily marked in large numbers, which undoubtedly cover much greater distances in their migrations. Many shore birds, such as the golden plover, nest in arctic regions and migrate in winter well south of the equator to take advantage of the second summer available in the Southern Hemi-

sphere. The arctic tern presents a famous case of
long-distance migration; it regularly travels from arctic to antarctic regions. This bird is a close relative
of the common tern, sometimes called "mackerel
gull" or "sea swallow," which is a familiar sight along
most ocean shore fronts and harbors in the summer
months. It has long narrow wings, and catches small
fish by plunging below the surface from several feet
in the air. Arctic terns nest along the coasts of north-

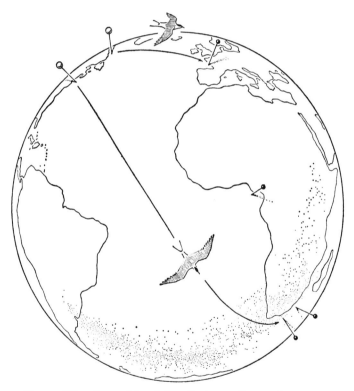

*Fig. 2. Migrations of banded arctic terns. Places where young
terns were banded off the coast of North America are shown by
pins with white heads. Points where these birds were recovered
a few months later are indicated by pins with small black heads.*

ern Asia and Europe as well as Canada, but in winter they are distributed widely over the southern Atlantic, Pacific, and Indian Oceans, as shown in Fig. 2.

Two spectacular recoveries of banded arctic terns demonstrate the potential speed and the extent of the migration in this species. Out of the large numbers banded on a small island off the coast of Labrador in July and August, a few have been retaken in France, West Africa, and South Africa. One bird attracted especial attention in 1928 when it was reported from the Indian Ocean coast of Natal, on the east side of South Africa, having flown around the Cape of Good Hope. This arctic tern was found dead only 116 days after it had been banded in Labrador on July 23 as a downy chick. Some years later another arctic tern banded in western Greenland on July 8 was retaken near Durban, Natal, on October 30 of the same year. Within a few weeks these two terns must have flown a minimum of 8500 and 9500 miles, respectively. If we take into consideration the rate at which tern chicks grow, it is clear that they could not have left their nesting areas for about a month after being banded. Making this correction leads to an estimate of about 100 miles per day as the average speed of their first fall migration halfway around the world. Furthermore, some of the arctic terns that migrate south along the eastern shores of the Atlantic Ocean from the arctic regions of both North America and Eurasia continue around the Cape of Good Hope well into the Indian Ocean. Another arctic tern banded near Murmansk on the arctic coast of the U.S.S.R. was recovered in Western Australia. To be sure, more than a year had elapsed in this case between banding and recapture, and we cannot be absolutely certain that the bird reached

Australia via Cape of Good Hope, nor that such a long migration occurs very often. Nevertheless these three extreme records demonstrate that the maximum extent of bird migrations is limited not by the capabilities of the birds, but by the size of the planet.

Since large-scale bird migrations occur every year, one might naturally inquire whether they can be observed directly. Perhaps the most obvious migrations are those of ducks and geese, which fly by day in the familiar V-shaped formations, often calling as they go. Few other birds migrate so conspicuously, however, and at any one moment it is difficult to tell whether a particular bird or group of birds is actually migrating or flying from one feeding place to another, returning to an area convenient for rest, or searching for new sources of food. Yet close observation by experienced students of bird behavior reveals subtle distinctions. Migrating birds are likely to show a persistence that is not evident at other times. Often they show less fear of men or natural enemies; for example, a hawk can seize one member of a migrating flock of small birds without arousing the panic ordinarily shown by the rest of the flock in such circumstances.

Migrating birds are sometimes concentrated into a relatively small area because of geographical factors, such as points of land that extend out into a body of water to be crossed, or mountain ridges that produce updrafts especially favorable for soaring. For example, Point Pelee, in southern Ontario, juts out some miles into Lake Erie, and on their fall migration many small birds gather there just before crossing the remaining several miles of water. Cape May, New Jersey, collects similar concentrations of small migrants. Hawk Mountain in central Pennsylvania serves as a convenient "moving sidewalk" for migrating hawks where they can ride the updrafts as the wind is deflected vertically by one of the folded Appalachian ridges along its many miles of length. But such situations are exceptional. Most of the time birds come and go, twice each year, without calling our attention to their actual migratory flights.

Nocturnal Flights

One reason why bird migrations are inconspicuous is that most of them take place at night. To be sure, there are many kinds of birds that usually migrate by day, and a rough and ready rule is that larger birds, and those that can feed on the wing while migrating, tend to travel long distances by day. Small and relatively secretive birds, on the other hand, tend to make their long migrations at night; this is especially true of insectivorous species that must seek out their food by intensive searching. Diurnal migrants include hawks, doves, swifts, swallows, crows, some seed-eating sparrows and finches, as well as herons, ducks, and geese. Groups of birds that customarily migrate at night are the flycatchers, orioles, together with most of the warblers, and other small perching birds. But there are many exceptions to these rules, and the very fact that diurnal migration is much more easily observed has produced a heavy bias in the available data. Even within one species both diurnal and nocturnal migration may occur at various times or places. As adequate methods have been developed to study nocturnal migration with greater precision, more and more birds have been found migrating under the cloak of darkness, including many waterfowls, shore birds, and terns, in addition to the small songbirds.

Birds migrating at night often emit characteristic

flight calls. Since some species use the same calls they employ in daytime, they can be identified by experienced ornithologists on nights of heavy migration. Others use different notes, and since the birds are too high to be seen in the darkness, it is often impossible to determine what kind of bird is calling. No one knows the function of these call notes, but it is a reasonable guess that they serve to keep individual birds in touch with one another, and in this way help to keep the flocks together. Usually, calling flocks are separated by considerable distances, and it is unlikely that birds in one flock can hear those of another group.

One of the most effective methods for studying nocturnal migration has been in use for almost one hundred years, and yet it has by no means been exploited to the full. It is simply to peer through a telescope at the full moon. Every few minutes on nights of heavy migration the patient observer can see a bird fly across the moon's surface. Sometimes the tempo of the wingbeats can tell the expert what general type of bird it is. Since the moon occupies only about half a degree of arc, or approximately 1/100,000th of the entire sky, a comparably small fraction of the birds flying overhead will happen to pass between the moon and the observer's telescope. Birds tend to fly within a few thousand feet of the ground, and more will cross the half-degree cone of visibility when the moon is low in the sky than when it is near the zenith. But even at best "moon watching" brings into view only a tiny fraction of the nocturnal migrants passing within a few miles of an observer. That the method works at all is eloquent testimony to the enormous numbers of birds that pass over most temperate regions during the nights of heavy migration.

One simple result of extensive moon watching is the conclusion that the maximum number of birds pass a given point in the middle of the night. An unexpected finding was that heavy migration was not always followed by a sudden influx or disappearance of migratory birds in the area of the observations. Much of the attention of bird watchers is drawn to concentrations of migratory birds that have just arrived at a given area or are just about to depart. These concentrations are more easily observed than the 0.001 per cent of the birds passing overhead that become visible through a telescope trained on the moon. But moon watching often shows a massive migration on a given night even though the local bird population remains unchanged the next morning. What is seen through the telescope is an enormous leap-frog movement, thousands of birds passing through the observer's area after starting far away and with still hundreds of miles ahead of them before morning. It may even be that arriving and departing birds fly too low to contribute much to the moon-watching counts.

Marathon Flight of Blue Geese

One especially favorable situation for tracing the whole of a long migratory flight developed in 1952 as a result of a combination of circumstances, including a collision between an airliner and a migrating goose. Blue geese, close relatives of the more familiar Canada geese, but a nearly uniform blue-gray in color, assemble every fall in the vicinity of James Bay, the southern arm of Hudson Bay, prior to a fall migration to the Gulf Coast of Louisiana and adjacent parts of Texas. On the evening of October 16, 1952, unusually large flocks of blue geese were observed taking off to the south from the mouth of the Kesagami River on James Bay. On the following day heavy flights of geese, probably including both blue and Canada geese, were seen by pilots of Trans-Canada Airlines at altitudes of 6000 to 8000 feet in the area just north of Lake Huron.

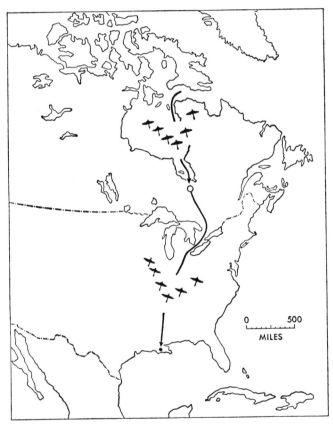

Fig. 3. Approximate route covered by flocks of blue geese that were observed leaving the southern end of James Bay (open circle) and arriving in Louisiana sixty hours later. Origin of this fall migration was somewhere in the vicinity of Hudson Straits, where this species nest.

One plane was damaged slightly by collision with a goose and forced to return to the airport at North Bay, Ontario. As a result, a warning was issued to all pilots in this area to watch for migrating geese. On October 18, large flocks of blue geese were seen flying south at about 3000 feet altitude over southern Illinois. Finally, on the morning of October 19, large flocks of blue geese arrived in Vermillion Parish, Louisiana. All these reports probably referred to the same migrating flocks, because blue geese are distinctly less abundant than Canada geese, and during this particular fall migration they were unusually concentrated. The main group of blue geese must have traveled 1700 miles along the route shown in Fig. 3, from the Kesagami River to Vermillion Parish, in approximately sixty hours, or at an average speed just under thirty miles per hour for the whole period of two and a half days. Since blue geese can fly at forty to forty-five miles per hour, they may have rested briefly along the way, or they may have remained in the air almost the whole sixty hours but deviated somewhat from the shortest airline route. Whatever assumptions one makes about their behavior between take-off and arrival, they clearly did not linger or waste much time in activities other than steady cross-country flight.

Migrations of Other Animals

It would be a mistake to infer from these descriptions of impressive flights of birds that annual migrations are unique to this particular group of animals. Obviously, the power of flight makes long migrations easier for birds than for animals with more limited powers of locomotion. Bats are flying mammals, and while they are not so conspicuous, so well understood, or nearly so popular as birds, they also migrate with considerable success. Seasonal observations of presence and absence have been followed up by the tagging of thousands of individual bats and recovery of a few at considerable distances. All bats living in temperate latitudes feed on insects, and in winter this food supply is almost totally lacking. Bats avoid starvation either by hibernating in some cool, but not subfreezing, shelter such as a cave, or by migrating south to where insect food is available even in winter. The latter solution was first suspected because some species of insectivorous bats were seldom if ever found hibernating in caves, even though very abundant in summer in Europe or the northern United States and Canada. Banding subsequently disclosed that even bats that do use caves as winter retreats commonly migrate as much as 150 or 200 miles from summer quarters to the particular cave chosen for winter shelter.

In the southwestern United States a particular kind of insectivorous bat, the Mexican free-tailed bat, is extremely abundant, especially in caves such as Carlsbad Cavern, New Mexico. Many thousands of these bats have been banded both in the southeastern states and in Mexico, and a number of banded bats have been retaken after migrating as far as 800 miles from New Mexico and Oklahoma to northern and central Mexico. In Europe, bat migrations of comparable length have also been demonstrated by banding, one of the longest documented flights having been from Dresden, in south central Germany, northeast to Lithuania.

Some mammals that cannot fly perform long migrations. The whales that are hunted for their meat and oil in the oceans surrounding the Antarctic Continent, have been marked by harmless, tagged harpoons, and some have later been killed hundreds of miles away. Seals are also known to migrate from breeding areas, such as the famous Pribilof Islands in the Bering Sea, south to tropical waters of the Pacific Ocean and back again. Even mammals that can only travel on foot, such as the bison of the American Great Plains, in former times, and the caribou of the arctic tundras, have seasonal migrations of 100 miles or more.

Sea turtles swim hundreds of miles between areas of the ocean where they feed, and certain beaches where they come to land to lay their eggs. For example, sea turtles that hatch from eggs on the beaches of islets near Ascension Island in mid-Atlantic, roughly halfway from Brazil to Africa, migrate west to the coast of South America and back again to this small group of islands. Other sea turtles migrate from beaches of Florida to distant parts of the Caribbean Sea. Even some of the small turtles of inland ponds and rivers migrate for a few hundred yards or even a mile or two from their usual habitats to sandy patches suitable for egg laying.

Many fishes also perform most impressive migrations, both in the ocean and in fresh water. Everyone has heard of the salmon, and other fish, that spend much of the year at sea, but swim up rivers in the spring to spawn in fresh water, swimming up waterfalls if necessary to reach their goal. Some salmon spend the majority of their time hundreds of miles away from the mouths of the rivers which they ascend so spectacularly to reach their spawning grounds. Many other fishes perform similar migrations from the deeper waters of lakes to spawning areas up small streams flowing into the lakes. Entirely within the oceans there are also extensive migrations, which have been more difficult to study because of the difficulty of catching or tracing free-swimming fish throughout the vast breadth and depths.

The ability to migrate is not even a monopoly of vertebrate animals. Squids migrate for at least 100 miles along the coast of Europe. On a small scale, many oceanic crustaceans migrate vertically every twenty-four hours from depths of several hundred feet up to the surface layers of the ocean. Usually they ascend at night, and thus remain in roughly the same, low level of illumination. But the best-known, and in many ways the most impressive, migrations of animals other than birds are found among insects. Many insects are so small, and such feeble fliers, that when they are carried long distances it is largely as passengers traveling along with a moving air mass. Some that hardly fly at all are lifted passively aloft on rising air currents and later dropped, gently because of their high ratio of air drag to body weight, hundreds of miles away. Even certain kinds of spiders travel in this passive fashion, except that they take the initiative of waiting for favorable conditions of winds and updrafts before secreting long trailing strands of silk that suffice to lift them from the ground. But quite apart from such free riders, many of the insects that are strong, fliers migrate actively over long distances, almost as much masters of their itineraries as bats or birds.

The Monarch Butterfly

Because insects are so small, so astronomically numerous, and so difficult to follow individually, insect migrations are known almost wholly from the simple and laborious recording of their seasonal presence and absence. This type of evidence has convincingly demonstrated migrations of many tens or hundreds of miles in a variety of butterflies, a few kinds of moths, and probably also in dragonflies. But recently the next logical method, tagging and numbering of individuals, has been applied to the monarch butterfly, an abundant and strongly migratory species. A monarch butterfly weighs only about 0.4 gram (one seventieth of an ounce), and is much smaller than the smallest hummingbird. When one watches even a large butterfly like this species in its flight, the first reaction is concern that at any moment it must falter to the ground. It seems almost incredible that such papery wings could carry so delicate an insect hundreds of miles in a regular, annual migration that is biologically advantageous to the species. For many years, the seasonal appearances and disappearances of monarchs and other butterflies had led biologists to the tentative conclusion that they must migrate north and south for considerable distances, despite the buffeting of winds, and the geographic barriers presented by rivers, lakes, or mountains.

Recently this conclusion has been confirmed and extended through a large-scale, co-operative tagging effort, led and co-ordinated by F. A. Urquhart of Toronto. The tags are small, numbered slips of paper glued to one wing, and many thousands of these butterflies have been tagged, for the most part by enthusiastic amateur biologists. Several monarch butterflies tagged near Toronto, Ontario, have been recovered as far east as Long Island, New York, and as far south as Florida and Texas. See Fig. 4. The most rapid journey over a very long distance was registered by one butterfly released in Ontario on September 13, and recovered in Texas, 1345 miles southwest, on October 25 of the same year. The minimum average speed for this migration was

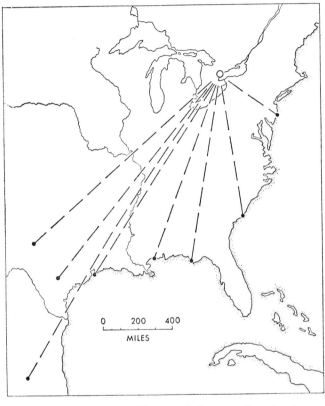

Fig. 4. Monarch butterflies, marked near Toronto (open circle), were recovered the following autumn at the points indicated by solid dots. The 1870-mile flight to Mexico was covered in four months, while one flight to Texas took place in 42 days.

thirty-two miles per day. The longest flight documented by recovery of a tagged monarch butterfly was from Ontario to the state of San Luis Potosi, Mexico. Four months and seven days had elapsed between tagging and recovery. No absolute speed record was set by this butterfly, but the airline distance traveled was an impressive 1870 miles.

We know far too little about the migrations of monarchs and other butterflies to add much of significance to the bare fact that they do cover distances comparable to those traveled by many migratory birds. But before we return to the more thoroughly studied migrations of birds, it is important to reflect upon the energetic resourcefulness required of a 0.4-gram butterfly when it sets out from Toronto and reaches the Gulf Coast a few weeks later. It often seems that our present knowledge of biology is limited to partial solutions of its simpler problems.

Chapter 2

BIRD WATCHING BY RADAR

A new method of studying bird migration recently has reinforced the tireless enthusiasm of bird-and-moon watchers. The story of how radar has come to be used to study birds is an interesting, and sometimes amusing, example of the interaction between biological and physical science. Radar was developed, of course, for military purposes, among which the detection and location of hostile airplanes were pre-eminent. After relatively primitive beginnings in the 1920s and 30s, all major powers involved in World War II devoted tremendous technological effort to improving their radar systems, so that they might detect enemy aircraft at greater ranges and plot the flight paths more accurately. A widely used method of presenting the information gathered by a radar system is the so-called Plan Position Indicator (abbreviated PPI) screen. This device is a specialized sort of picture tube, like that of a television set. Inside this picture tube a beam of electrons accelerated through a vacuum bombards the fluorescent material on the inside of the round face of the PPI tube. The motions of this electron beam, and variations in its intensity, are automatically manipulated by the radar set to draw on the face of the picture tube a map of the area around the radar installation. Compass directions are accurately represented, and the distance of an airplane or other target returning radar echoes is shown to scale by the radial distance of a "blip" of light from the center of the screen.

In the PPI system a narrow beam of high-frequency radio waves sweeps horizontally around the points of the compass, while at the same time the beam of electrons inside the picture tube follows an exactly parallel course around and around the circular radar screen. As this electron beam rotates in exact step with the broadcasting antenna overhead, it is deflected radially inward and outward so that it traces successive radii of the circular screen itself. If no echo is received, the intensity of the beam is kept low enough that it makes no conspicuous mark. But when an echo of the emitted radar pulse does reach the receiver, it causes a momentary intensification of the electron beam and the resulting spot of light, as shown in Fig. 5. The rate of motion of the beam in and out along a given radius is precisely related to the time taken by the radio waves to travel out to a given distance and back again. For example, a given radar set might be adjusted so that the radial distance from center to edge of the screen corresponds to twenty-five miles of round-trip travel by the radio waves, at the speed of light, approximately 3×10^{10} centimeters per second. Thus twenty-five miles corresponds to a round-trip travel time of about 2.7×10^{-4} second, and the radial motion of the radar beam is so adjusted that it travels out from center to periphery of the screen in this same, minute interval of time. Each object producing a radar echo makes its mark on the map thus sketched by the electron beam on a fluorescent ground-glass face of the picture tube. (For a more

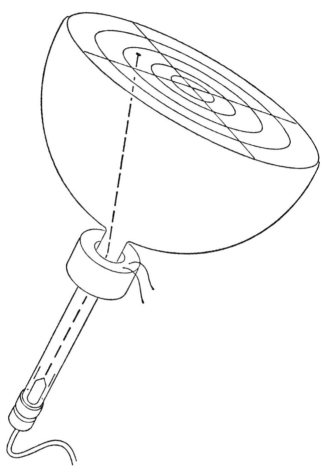

Fig. 5. *Simplified diagram of radar PPI tube in which a beam of electrons is represented by the dashed line. Plates and coils deflect the beam to make it strike the tube face at various positions and thus to draw a map.*

complete discussion of radar see *The Origin of Radar*, by Robert M. Page, Doubleday Science Study Series, 1962.)

The Limitations and Imperfections of Radar

Like all mechanisms, a radar set and its PPI screen have their limitations and their imperfections. Very strong echoes make bright and unequivocal marks, but smaller or more distant targets register their presence more faintly. The amplification of the set can be increased, but eventually the random agitation of electrons in the circuit itself, or the scattering of radar waves from the atmosphere, produces a background noise or "snow" that obscures the echoes from small and distant targets. In seeking the greatest possible sensitivity, radar operators adjust the amplification of their systems to bring out the faintest practicable targets. This often requires adjusting the amplification to a point just below the level where the whole screen becomes filled with the random lights originating as noise.

As radar experience accumulated, interfering "noise" from the atmosphere became a constant problem, and a serious limitation to the effectiveness of the radar for its intended purpose. When radar beams became more powerful, and receivers more sensitive, there began to appear from time to time a variety of spurious indications that something other than aircraft was "out there." Rain, snow, and even the microscopic droplets of water or the ice crystals that make up clouds can all produce radar echoes, as may dust storms and swarms of locusts. Clouds and rain or other forms of atmospheric water yielded characteristic radar patterns. Thunderstorms and hail could be recognized, and severe storms, including waterspouts and tornadoes could be detected in time to broadcast warnings of their approach. In fact, special radars have been developed for the use of meteorologists.

But other nuisance traces continued to plague the radar engineers, and some of these did not remain in one place, but moved about. They often appeared and disappeared without rhyme or reason, and in mystified exasperation radar operators dubbed them "angels." The choice of this term, rather than one carrying an implication of diabolical intent to interfere with the radar, may be taken as a touching tribute to the faith of technical people in the fundamental goodness of nature. All through World War II every detail of radar operations was a deep military secret, especially information about defects of particular systems that might be of value to an enemy intent upon eluding radar detection. Consequently, the taxonomy of radar angels, instead of being exposed to public scrutiny and debate, remained for years the subject of confidential committee meetings and "classified" mimeographed reports. A variety of puzzling types of angels appeared consistently, and since they tended to occur at certain seasons and particular times of day, the engineers concerned with improving radar effectiveness looked for explanations in meteorological phenomena.

One clear example was provided by the "ring angels" noticed in certain parts of England. Suddenly an otherwise blank radar screen would show a distinct spot at a certain point on the map. The spot would increase in intensity and size and then, by gradually fading at its center, turn into a ring. The ring would grow in size but weaken in intensity until within a few minutes it disappeared. Some of these ring angels were displayed rather regularly day

after day at the same point. Were they some unknown type of thermal air current rising under the
heat of the morning sun, perhaps carrying aloft invisible dust particles that gave radar echoes? Or was
some other natural phenomenon at work (or at
play)?

Radar Angels

Suggestions were advanced that birds might be
reflecting radar waves, but for some time this notion was dismissed as foolish nonsense. Everyone
knew that birds' bodies were soft and nonmetallic
and, besides, there weren't that many birds around.
But slowly the evidence accumulated that radar angels shared many of their properties with birds. They
moved at about the correct speed, and they moved
upwind, which eliminated clouds and particles suspended in the air. They were most abundant in
spring and fall, and they were limited to altitudes of
a few thousand feet, which excluded electrical phenomena in the upper atmosphere. Finally, someone
investigated the places where ring angels appeared
most consistently, and found that they coincided
with very large starling roosts. The strongest ring
echoes occurred just when thousands of starlings
flew up from their nighttime roosts and scattered in
all directions in their daily search for food. Swayed
by these facts, some radar engineers began to
mount telescopes on their radar antennas, and when
angels occurred under conditions favorable for visual
observations, they found that birds could be seen at
the distances registered for the angels on their PPI
screens. There are still unexplained types of radar
echoes, but a large fraction of the angel population
turned out to have been flying on feathered wings.

Once it was recognized that the diffuse milky angels on radar screens were in fact birds, it became possible to use them for investigations of migration. But this new method has so far been limited, unfortunately, by the wholly reasonable preoccupation of the best radar systems with military enterprises. Only occasionally, and as a very incidental by-product of their basic purpose, can the enormously expensive radar installations be diverted to study bird echoes. It is possible, and ordinarily desirable, to adjust a radar set so that it minimizes angels and brings out airplane echoes instead, but of course this renders the radar almost useless for biology. One procedure that has proved useful, though tantalizingly inadequate in comparison to what would be possible under ideal conditions, is to have the military crews who operate our powerful radar installations take motion pictures of the PPI screens when bird echoes are prominent, and allow interested biologists to analyze these photographs. Much more could be learned about the steady flow of birds passing over the several hundred square miles surrounding each major radar system, if the radar beams could be trained specifically on interesting flocks of migrants, and the installations operated with the schedules of bird migration in mind, rather than other needs. In due course, when present radar systems have been rendered obsolete by newer ones, such specialized biological uses may become feasible.

Careful study reveals that not all angels are alike. Sometimes the white spots are faint, sometimes so large and bright they fill the whole screen (like those shown in Fig. 6). Usually the radar beam rotates horizontally at one revolution every several seconds, and since each speck of bird echo has

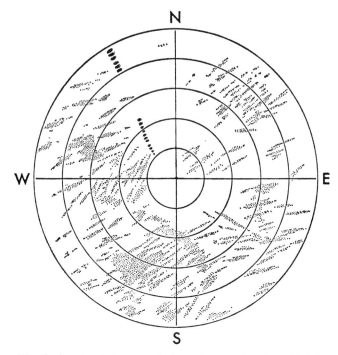

Fig. 6. A radar PPI screen during a heavy migration of birds. Echoes from flocks of birds produce the diffuse echoes; airplanes generate dense concentrated blips like those in the upper left-hand quadrant. The airplanes have moved appreciably between circular sweeps of the radar antenna, but the birds moved so slowly that the echoes fuse.

moved in that interval, the whole pattern is one of constant flux. When motion pictures are made up from pictures of the PPI screen taken about once every 10 or 12 seconds, and projected at 16 frames per second, a whole night's migration may be telescoped into a few minutes. The net impression is similar to what a man might see if he watched drifting patches of mist through a narrow round window marked out in concentric rings like a gunsight.

Each speck is not a single bird, but a trace formed
on the fluorescent radar screen by some interaction
between hundreds of radar pulses and dozens of
birds. The resulting electric currents flow through
complex electronic circuits, are amplified and caused
to deflect many millions of electrons before they ac-
tivate fluorescent molecules lining a ground-glass
screen. No one has yet seriously attempted to un-
ravel the relationships between these shifting glints
of light, comparable to the snow on a poorly adjusted television set, and the actual population of
flying birds. Often there must be several kinds of
birds of different sizes, beating their wings at different rates, flying at somewhat different altitudes,
and perhaps not all in quite the same direction. One
speck on the PPI screen corresponds to the birds
flying through thousands of cubic feet of air, several
miles away. The wonder is that radar reveals anything
at all about the birds, not that it tells so little.

Another puzzle of radar bird watching is that bird
echoes may disappear altogether from time to time,
even when it is unlikely that the birds have landed.
It may be that only concentrated flocks of small
birds return strong echoes; perhaps if the flocks are
dispersed the same over-all density of migrants produces fainter echoes, or none at all that rise above
some background noise level. The picture of bird
migration obtained by radar may be distorted by un-
due emphasis on populations of birds, especially
concentrated groups, rather than on widely dispersed individuals.

Despite these complications and limitations, the
experienced students of birds and their radar echoes
have begun to classify and hesitantly to identify
categories of bird echoes. For example, William
Drury, Jr., of the Massachusetts Audubon Society,

and his colleagues, J. Baird, J. A. Keith, and I. C. T. Nisbet, distinguished the following types of "angels." First there are large intense spots that move slowly and represent gulls (in the area of Cape Cod, mostly herring gulls). In other radar studies, especially those of Ernst Sutter at Zurich, Switzerland, the gulls can be recognized every morning and evening as they come and go between Lake Constance and the Zurich city dump. Then there are intense fast-moving angels visible at long range which represent shore birds or waders (sandpipers, plovers, and similar birds). Finally, there are small, weak, and diffuse spots which come and go on the screen, although when present they maintain a steady migratory movement. These are almost certainly small songbirds, especially warblers.

If these uncertainties of identification appear disappointing, one should recall that the registration of songbird echoes at twenty miles or more was a total surprise, wholly unexpected by the ingenious designers of the radar systems. Even while used to study bird migrations merely on a "so as not to interfere" basis, radar has contributed important confirmation of many points previously inferred from indirect evidence. It has extended our picture in many ways, most importantly by demonstrating the amazing extent of the mass movements of birds, and the readiness of even the smallest of birds to strike out across stretches of open oceans. For example, the analyses of bird echoes departing in the autumn from eastern Massachusetts, show frequent mass flights of small songbirds southeast or south over the ocean. These warblers and other birds of similar size move at twenty-five to thirty miles per hour as far out over the Atlantic as the radar can detect their echoes. The blackpoll warbler is one of the most common species following this migration route and contributing to the radar echoes that move south to southeast from Massachusetts in the general direction of Venezuela. When the radar shows heavy southerly flights of small songbirds at twenty-five to thirty miles per hour, bird watchers are very likely to find a marked decrease in the population of blackpoll warblers on the following day. On certain occasions of this sort an influx of blackpoll warblers into Bermuda was noted on the following day. But so many thousands of small birds are "seen" by the radar systems, flying south and southeast from Massachusetts, that only a small fraction can be accounted for by those that stop over for a few days in Bermuda. The great majority must continue the full distance of 2000–2500 miles to the Antilles or South America.

In the last chapter were summarized the most rapid records of migratory flights demonstrated by the recovery of banded birds, and most of these indicated average flight speeds of one to three hundred miles per day. It was pointed out that these were minimum speeds, since it was impossible to determine how much time between banding and recovery the bird spent in other activities than migratory flights. The radar analyses show that even small birds migrate steadily at their normal flight speed. The routes along which they start, and which they maintain across the oceans, demonstrate that many must fly for at least sixty to seventy hours without pause. Seventy hours at thirty miles per hour will take a blackpoll warbler 2100 miles, neglecting any effects of wind. These birds are known to travel non-stop at least as far as the 1800 miles from Massachusetts to Puerto Rico, and it is very likely that many of them continue to the Lesser Antilles or the mainland of South America without landing. From observational evidence and recoveries of banded birds, biologists had concluded long ago that such flights must occur, but the radar echoes have shown them to be routine activities of millions of birds every autumn.

Altitudes at Which Birds Migrate

Even in the past sixty years, when airplanes have been carrying men aloft in controlled and rapid flight, the altitudes attained by flying birds have remained impressive, if only because these altitudes are maintained for hours or days solely by the work of living muscle. The older observations of altitudes of bird flight had of necessity to be made from the ground, but even those calculated estimates demonstrated that many birds can fly efficiently at heights where the air contains appreciably less oxygen than in our familiar, narrow zone near sea level. Olympic games are not held, nor are international records set, at 5000 feet, where the oxygen content of the air is about 20 per cent less than at sea level. But rooks have been observed flying contentedly at 11,000 feet where the oxygen content of the air is only 66 per cent of that at sea level, and ducks and plovers have been watched from airplanes while migrating at approximately normal speed at 7500 to 8500 feet. Shore birds such as curlews, and also cranes, have been seen migrating near Mt. Everest at altitudes which the mountain climbers could be sure from their own position were above 20,000 feet. Geese have been observed migrating over the Himalayas at an estimated 29,500 feet. At this altitude the air contains only about 30 per cent of the sea-level oxygen concentration.

When nocturnal migrants are seen through telescopes against the moon, it is impossible to estimate their altitude with significant accuracy. Certain types of radar systems, however, give excellent information about the heights reached by flocks of birds. An extensive analysis of bird echoes or "angels" photographed from the PPI screen of a powerful radar installation on Cape Cod included records from a radar heightfinder. In this special application the radar beam is moved vertically up and down and radar echoes are automatically registered on a screen in the form of a graph of altitude above the ground as a function of distance from the radar installation. The horizontal direction in which the vertically scanning beam is pointed can be varied at will by the operator. One limitation of the system is that radar echoes from the ground or water confuse the record below altitudes of a few hundred feet, so that only relatively high fliers are represented in the statistical sample analyzed by this type of radar system. In the most thorough radar analysis yet available, covering observations over forty-five nights during seasons of heavy migration, it was found that the most common altitude of migration for all types of birds combined was 1500 to 2500 feet above the ocean. On the average only 10 per cent of the bird echoes came from above 5000 feet, and less than 1 per cent from higher than 10,000 feet. These average altitudes of migration were not appreciably different when the radar beam was pointed out to sea or when it was scanning the air over the mainland some miles inland. Other studies employing radar, and occasional encounters between migrating birds and airplanes, have yielded a picture quite consistent with these radar data from Cape Cod. Radar observations in England, using a somewhat different type of radar system, showed many birds up to 5000 feet, and a few, especially on clear nights, up to 13,000 feet.

While relatively few migrants fly at very high altitudes, those few are of considerable interest, because necessarily they have succeeded in overcoming the special physiological problems of muscular exertion in the thin air of high altitudes. The radar observations of Cape Cod showed some echoes from altitudes up to 20,000 feet or more. A few appeared from their detailed appearance and speed of movement to come from small songbirds, but most of these truly high fliers were shore birds such as sandpipers and plovers—as demonstrated by their speed of flight. Radar data are not yet available from mountainous areas, but direct observations such as those mentioned bear out that massive migrations occur over the tops of the highest mountain masses. The most impressive are those of birds that migrate twice each year over the Himalayas from central Russia to India. Sometimes such migrants appear to be concentrated in the passes, but in the Himalayas even the passes are mainly above 10,000 feet. Furthermore, in good weather many flocks of migrants appear to ignore valleys and climb to altitudes where they can clear all but the very highest peaks.

At about 18,000 feet the air contains only half as much oxygen as at sea level. Human mountain climbers, depending on their degree of acclimatization and physical stamina, begin to find heavy exercise difficult between 10,000 and 20,000 feet. As every newspaper reader knows, climbing the truly high mountains of the world, in particular Mt. Everest, has been possible only by the very best of mountaineers, with massive expeditionary support, and with tanks of oxygen to breathe at frequent in-

tervals. One need only read the accounts of John
Hunt or Edmund Hilary, especially their vivid descriptions of the painful and exhausting effort required merely to crawl out of a sleeping bag and put
on one's boots, to appreciate what it must mean for
geese to fly on their own power over the Himalayas
at an estimated altitude of 29,500 feet. This report
of migrating geese flying higher than the summit of
Mt. Everest is an isolated one, but many Himalayan
expeditions have reported migrants and resident
birds flying without apparent strain at altitudes
where even acclimated mountain climbers must
rest every few hundred feet of travel on foot. No one
knows what the metabolic rate of a goose may be as
it migrates over the Himalayas, but these flights
must require a strenuous and sustained effort. Human mountain climbers must acclimate themselves
gradually over some weeks before they can struggle
(and then with great difficulty) to heights above
20,000 feet, but migrating geese probably take off
from the plains of Siberia and fly within a single
day up to these altitudes and down again to the rivers and lakes of India. This striking biological phenomenon has not been studied enough to provide
any hints as to how birds avoid altitude sickness,
or how they muster enough muscular energy to fly
in air containing only about one quarter of the oxygen available at sea level. Any explanation of this
impressive accomplishment must wait for some
future investigator to devise adequate methods for
studying the energetics of bird flight at high altitudes.

Speeds of Migratory Flight

For more than one hundred years those interested
in birds have made estimates of the velocity of bird

flight, but it is surprising how difficult it is to obtain representative and truly accurate figures. One of the earliest errors was to neglect the difference between ground speed and air speed—and even to make the naïve assumption that they were identical. In these days of commonplace travel by airplane everyone realizes that the air is moving, and that a bird or airplane travels over the ground at some velocity that is the net resultant of its speed through the air and the wind velocity. What must be added to this simple consideration is the fact that the air moves at very different speeds, and sometimes in different directions, at various altitudes. Usually winds are stronger at several hundred feet elevation, because the ground causes some friction, turbulence, and a general reduction in air flow at low altitudes. Thus, when the ground speed of a bird flying a few hundred or thousand feet above the surface has been measured or estimated, it does not suffice to make a correction solely on the basis of the wind velocity at ground level. Meteorologists at major airports measure the "winds aloft" by means of balloons that are tracked either visually with specialized telescopes that measure angular positions of the balloon as a function of time while it rises at a known rate, or else by radar. Another method is to fly close to birds in an airplane, and either adjust the plane's flying speed (measured by the standard air-speed meter) until it approximates that of the bird or else estimate in some way how much more slowly the bird is moving. With the most rapid fliers the former method has given fairly accurate values of air speed, but there is always some question whether the proximity of the airplane did not spur the bird to greater speed than normal.

Some years ago Colonel Meinertzhagen of the British Army made a thorough study of the flight speeds of birds, employing these methods and critically evaluating previously published reports of birds' flight speeds. His most reliable maximum values for air speed in level flight of the more rapid fliers include, mallard ducks at 60 m.p.h., swifts, golden plover, and hummingbirds at about the same speed (57 to 62 m.p.h.) despite their great disparities in size and type of flight. Many birds in steep dives exceed 100 m.p.h. There is one recorded instance when an airplane flying at about 100 m.p.h. was overtaken and passed by a flock of sandpipers or small plovers that must have been traveling at roughly 110 m.p.h. But the smaller songbirds fly most of the time at speeds on the order of 30 m.p.h.

Here again radar has given more reliable estimates of typical ground speeds attained by flocks of migrating birds. The radar echoes studied by Drury and his collaborators, and identified as being from shore birds such as sandpipers and plovers because of their compactness and high altitude, moved about 45 m.p.h. steadily and for hours on end. Probably Meinertzhagen's maximum figures of about 60 m.p.h. were for unusual bursts of speed in special circumstances, rather than rates that can be maintained steadily during long migrations. Radar echoes associated with small songbirds travel more slowly, typically about 30 m.p.h.

Directions of Migratory Flights

The simple statement that migratory birds fly north in spring and south in autumn is a considerable oversimplification, even though this is as nearly correct as any brief summary statement could be. Rather trivial exceptions occur where birds make short altitudinal migrations from mountaintops to lowlands, and in tropical rain forests, where there is little long-distance migration, there are nevertheless predictable, seasonal shifts of population from one source of food to another during the rainy and dry seasons. The migrations of oceanic birds may occur in almost any direction. For instance, the white-fronted terns that nest in New Zealand migrate almost due west in the fall to winter along the eastern and southeastern shores of Australia. Red-headed ducks that nest at the edges of small ponds in our own prairie states migrate almost straight east to winter along the Atlantic Coast. The rose-colored starling that winters in India flies only slightly north of west to nest in the steppes of Turkey and southern Russia. In the Southern Hemisphere one finds the expected reversal of compass direction as birds move north toward the tropics in the southern autumn. Fewer land birds than one might expect have been definitely shown to undertake such migrations; the explanation may be partly that there is so much less land area in southern temperate latitudes by comparison with the Northern Hemisphere, and partly that the birds there have been studied much less intensively.

Sea birds in the southern oceans undertake extensive migrations, especially those that nest along the shores and coastal islands of the Antarctic Continent. Even the penguins, which must do all their traveling by swimming, migrate hundreds of miles southward to Antarctica in the spring, and back to warmer waters and coasts during the winter. One of the best studied annual migration cycles of any truly pelagic bird is that of the Wilson's petrel. This is one of many species of storm petrels, about the size of an American robin, which are familiar to sailors as Mother Carey's chickens. They ordinarily fly just above the surface of the waves, often paddling their webbed feet against the water surface as an apparent aid to flight. Though they seem feeble and almost mothlike at a brief glance from shipboard, they ride out ocean storms without difficulty and spend most of their lives far from land. The Wilson's petrels nest on small islands between southern South America and the adjacent part of the Antarctic Continent, especially South Georgia, the South Shetlands, and South Orkneys (so named by European explorers who likened them to familiar island groups north of the British Isles). At the end of March and April, which of course is autumn in the Southern Hemisphere, these petrels migrate north to the tropical waters of the South Atlantic. By June they are very abundant off the east coast of the United States, much more so than the native Leach's petrels, which at that season are nesting on islands off the coasts of Maine and Nova Scotia. Later in the northern summer they tend to be over the North Atlantic, from New England and the Gulf of St. Lawrence to the Bay of Biscay and the northwest coast of Africa. By October they are migrating back southwards, and are observed from tropical waters of the Atlantic, especially off the coast of Africa, south to the vicinity of their nesting colonies. As shown in Fig. 7, despite their small size, and relatively slow flight, these petrels migrate almost as far as the arctic terns, whose exploits we already have discussed with admiration.

The majority of Northern Hemisphere migrants do move generally north and south, but their actual flight paths commonly deviate 45 to 60 degrees from due north in spring or south in fall. Some of this deviation results from following coastlines or river valleys, but not all. In western Europe many birds fly southwest in fall from Scandinavia, England, or northern Germany to southern France or Spain. Others from the same nesting area of northern Europe fly in a southeasterly direction to the Near East in fall. Many of both groups stop in the Mediterra-

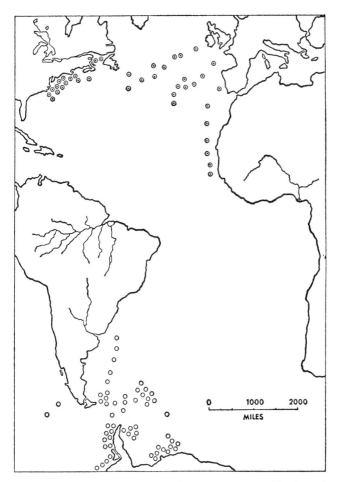

Fig. 7. *Observations of Wilson's petrels in the North and South Atlantic. Open circles are points where these birds have been reported during the month of January; circles with dots are records of July sightings.*

nean area, but at least as many other species continue south into Africa. In the annual migrations of
hundreds of species and millions of individual birds
from Europe into Africa there is a tendency to avoid
the long sea crossing by flying around either the
eastern end of the Mediterranean Sea, or through
Spain and across the Straits of Gibraltar. On the
other hand, there are many species that do cross the
Mediterranean without such deviations. A clear example from the former group that tend to detour
is the large, conspicuous, and well-known white
stork. Many thousands of storks have been banded
in western Europe, and recoveries along their whole
migration route detail approximately how they
travel, as shown in Fig. 8.

Very similar general statements can be made about
bird migration in North America, substituting for the
Mediterranean Sea the Gulf of Mexico. The major
difference is that the extensive moutain ranges of
the western United States and Central America tend
to keep the Pacific-coast species somewhat separated
from the birds that nest in the central plains and
east coast. Even this statement has its exceptions,
however, especially among waterfowl. Many North
American birds migrate more or less southeast from
wide breeding ranges in the eastern and central
United States or southern Canada to the area of
Florida, the Bahamas, and the eastern islands of the
Caribbean. Some follow this route from farther
north, among them the shore birds that nest along
the Arctic Ocean in northern Canada, and they
reach the Atlantic near the mouth of the St. Lawrence River. Many of these shore birds continue on
southeast or south into South America. The winter-
ing areas of certain species are far south the
equator, for example, the bobolink, or the golden

Fig. 8. *Fall migration of the European white stork. Storks nesting in France and western Germany migrate through Spain, while those nesting farther east pass around the eastern end of the Mediterranean.*

plover (see Fig. 9). Many migrants do not hold to a single compass direction throughout an entire spring or fall migration. The maps of many migration routes show distinct turns that occur at certain points in the journey; for example, northern European birds that fly around the Mediterranean shift from a southeasterly or southwesterly course to a more directly southward heading when they reach Morocco or Egypt.

Age and the Migratory Tendency

A further remarkable feature of the migrations of many species is the fact that birds of different age groups seem to migrate separately. Especially during fall migration the relative number of adults and young are quite different at various stopover points. In other species, the young birds seem to have a stronger migratory tendency, and are found farther south in winter than the adults. The common herring gull of the northeastern United States and eastern Canada is often found in winter throughout much of its summer range. But the gulls that remain all winter in colder latitudes are mostly adults, easily distinguished from birds up to three years of age by their "cleaner" gray and white plumage. When thousands of young herring gulls were banded at nesting colonies on islands off the coast of New England, the great majority of the recoveries at a considerable distance south (southern United States and the Caribbean) were of birds in their first winter. In other species, especially herons, the young birds scatter in almost any direction during the late summer and early fall of their first year. When these birds have been banded in large numbers as nestlings, the recoveries within the following few months are as likely to be to the north as south.

In other species the sequence may be reversed; the adult birds depart earlier than the young, which are thus left to their own devices in the same region where they hatched from their eggs. One of the extreme cases of this type is found in the petrels and shearwaters. These birds spend almost their entire lives flying over the open ocean, feeding on small fish and crustaceans which they can catch at the surface. Only in the nesting season do they come to land, where most species dig burrows to shelter their nests and eggs from predatory birds such as gulls. The young chicks hatch in the dark burrows and are fed by their parents, which usually take turns at brooding, one spending three or four days in the burrow incubating the eggs, while the mate is at sea, feeding. The chicks are then fed partly digested seafood regurgitated by the parents. Unappetizing as this diet may seem to us, it fattens young petrels until at a few weeks of age they are heavier than their hard-working parents. They grow so obese and ungainly that they cannot squirm out through the tunnel through which the parent birds enter and leave. In late summer the petrel squabs are deserted by their parents, and they fast for some time while their flight feathers grow out and their shape recedes to the streamlined, adult proportions. Only then can they leave the burrow and head out to sea to begin a wholly new life in a world with which they have had no experience whatsoever. Ordinarily there are no adult petrels in the area, and all the behavior of the young birds, including finding the ocean, finding food, and migrating over the sea for some hundreds of miles, must be accomplished on their own with only the guiding information contained within their own nervous systems.

Another bird of especial interest in this connection is the American golden plover, a moderately large shore bird that nests in arctic Canada. In winter the adult golden plover fly southeast, as mentioned previously, until they turn south in the general area of Nova Scotia and Newfoundland. After

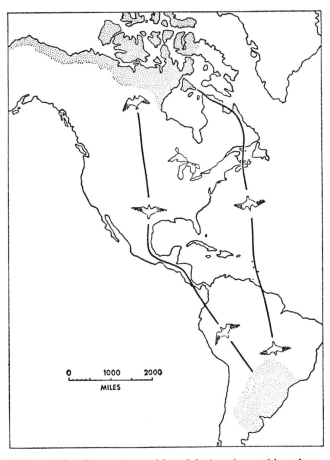

Fig. 9. *Migration routes used by adult American golden plover in fall and spring. Young birds of this species usually fly south along the same general route used by the adults for their northward migration in the spring. They thus find their way from the arctic regions where they were hatched to a winter range in South America without the guidance of adults of their species.*

an overwater flight of roughly 2000 miles they continue south, as shown in Fig. 9, across the Amazon Basin to temperate latitudes in Argentina, Uruguay, and southern Brazil. Here they find suitable habitats where they can feed on small invertebrate animals exposed at low tide on sand and mud flats—much as they do in the arctic or during their migrations. In spring, however, the golden plover do not retrace the same route at all, but fly northwest over the center of South America, through Central America, and north over the Great Plains and the valley of the Peace and MacKenzie rivers, back to the arctic coast. To add one further surprise, it has been found that almost all the young golden plover take this inland route south in the fall, while the great majority of the adults fly the Atlantic Ocean route outlined above. Like the recently slenderized young petrels leaving their burrows for the first time in their lives, the young golden plovers, deserted by their parents, start south for a journey about 8000 miles in length, entirely on their own.

Chapter 3

THE SEASONAL TIMING AND ENERGETICS OF MIGRATIONS

Long migrations are arduous and risky; bad weather is often the immediate cause of heavy mortality. Small birds expend a major portion of their bodily fat reserves in a single night of migratory flight. If, by some misadventure of crosswinds, morning finds them far at sea, exhaustion and drowning become imminent dangers. Many of the smaller migrants that fly aboard ships at sea are in a desperate situation. Another hazard of spring migration is to arrive too early and find the breeding area still snowbound or devoid of suitable food. Insectivorous birds, such as swallows, are especially vulnerable to spells of cold or rainy weather when they first arrive in the north. Without insects to catch, they starve, because their bills and digestive systems are too specialized for feeding on insects to permit subsisting on seeds or other available foods, even in an emergency. Hazardous at best, long-distance migrations would be suicidal without appropriate seasonal timing.

The Advantages of Migration

Since food and all necessities of life are available to birds all year long in the tropics, why do they subject themselves to these multiple risks? If migrations were not advantageous, natural selection would have tended to eliminate them. As a matter of fact, there are a few species closely related to migrants which do remain in much the same area throughout the year. If such a life is possible for one species, why not for its close relatives? The customary, though slightly evasive, answer biologists give is to point out that each species exploits one possible way of life. It is evident that both migratory and nonmigratory species of birds have succeeded well enough to have survived as members of the current fauna. If, however, migration conferred a major handicap, one would expect nonmigratory species to increase at the expense of globe-trotters. But quite the opposite has happened; migration has become the rule rather than the exception.

One advantage of a summer in the far north is that a large area is temporarily available to the migrants. Space for establishing nesting territories is plentiful, and while predators are always present, the nesting birds can scatter widely and reduce the chances that their nests will be discovered. If the summer temperatures are much cooler than farther south, they are at least tolerable. Few birds seem to be killed by cold itself, even under winter conditions in the north. Another advantage gained by migrating to high latitudes (which is not apparent at first thought) is the marked increase in the hours of daylight, culminating above the Arctic Circle in continuous light throughout the height of the summer season. Most birds are active only by day, and longer days therefore mean more hours available for whatever activities may be important to them. Raising young is clearly one of the most critical phases of the entire life cycle. Eggs and nestlings are tempt-

ing and vulnerable sources of food for many sorts
of predatory animals, even including those that cannot make a successful attack on adult birds in good
health and capable of flight. The development of
each young bird in its egg is speeded by incubation
at the relatively high body temperature of the parent,
and since birds maintain a nearly uniform body temperature, the development of the embryos does not
proceed at significantly different rates in the arctic
or tropics. In the far north the possibility of pro-
longing the daily feeding process by several hours,
beyond what would be possible in the limited daylight of temperate or tropical latitudes, may serve to
speed up the nestling phase.

Robins in Umiat, Alaska

Having suggested a hypothesis such as this, we
often find it unexpectedly difficult to test its correctness. The most direct question to be asked is
whether or not birds nesting in the continuous daylight of the arctic summer do actually bring their
young through the fledgling period more quickly than
others of the same species nesting where the days are
shorter. Other factors may complicate the picture—
for example, the availability of food. In only one,
very limited case do we have directly comparable
measurements at high and low latitudes of the time
spent as vulnerable nestlings; that is, the time from
hatching until the young birds can fly by themselves
and leave the nest. This comparison involved one
family of the common eastern robin which Martin
Karplus found nesting at Umiat, Alaska at 69° 23'
north latitude in late June when the sun varied in altitude from 2½° above the horizon at midnight to
46° at noon. From their fourth to sixth day after

hatching the four nestling robins and their parents were observed continuously from a blind nearby in a thicket of low willows. Karplus enlisted me and two other companions to share the forty-eight-hour watch in shifts, and we noted the times when the parents arrived with food, when the nestlings were brooded, and when the mother slept on the nest during the coldest hours near midnight. At this time the light intensity was minimal, though adequate for reading.

Fortunately, similar studies of the same species had been carried out in Ohio, and the progress of Karplus' arctic robins could be compared directly with the rate of development of young robins in a temperate climate. While nestling robins in Ohio were fed by their parents approximately every ten minutes for sixteen hours each day, our Alaskan robins received food at about the same rate for twenty-one hours. The average number of feeding visits per day was ninety-six in Ohio and 137 in Alaska. The continuous observation of the robin family at Umiat was stopped after the sixth day, partly to avoid disturbance that might cause the young birds to leave the nest prematurely. But we continued to inspect the nest from a distance, with binoculars, twice a day until the young flew off on their ninth day after hatching, normal in size and feather development.

Out of eighty-five families on which reports were available from the northern United States, the average length of the nestling period was 13.2 days, and only one brood left their nest as early as the ninth day. This one nine-day nestling period and most of the ten-day periods in the United States were cases where the birds had been disturbed sufficiently to cause premature departure from the nest. It thus seems likely that the development of the Umiat

robins to the flying stage, in nine days instead of
thirteen, represented a true acceleration of development by the longer period of parental feeding. It
was unfortunately not practicable to extend this sort
of observation to other robins' nests in the arctic,
partly because the eastern robin is a relatively scarce
bird this far north. Despite the numbers of biologists, amateur and professional, who study birds and
their migrations, no one has repeated this sort of
comparison of the same species nesting in arctic and
temperate latitudes. One brood of robins is not
enough to settle such a question as this, any more
than one swallow suffices to establish spring's arrival.
Perhaps these pioneering robins at Umiat fed their
chicks so well because of the same exceptional en-
terprise that brought them to arctic Alaska in the
first place. But in the absence of other information,
it seems reasonable to accept Karplus' interpretation
that a major advantage gained by birds from their
lengthy migrations is the more rapid development
of the nestlings through that vulnerable period when
they are succulent but flightless.

Readiness for Breeding

Since many of the advantages to be gained from
a northern summer relate to the rearing of young,
it is scarcely surprising to find that seasonal fluctua-
tions of the functional readiness of the reproductive
system are closely correlated with the annual cycle
of migrations. In some birds the plumage of the
males develops bright and conspicuous colors during
the breeding season. On the other hand there are a
few specialized species in which the situation is reversed. The female assumes the brilliant plumage
and relegates to the dull-colored male the tending of

eggs and young, as in phalaropes (and certain modern women).

More closely related to the progressive preparations for producing young, though not visible externally, are the relatively great changes in size and activity of the ovary and testes. It is not at all unusual for a tenfold change in weight or volume to occur in these organs where the eggs and sperm are actually produced. In the female the oviduct and its accessory glands that produce the white of the final egg, and its shell, also fluctuate greatly in size and state of development. It is thus very easy to tell on dissecting a bird whether or not it is in breeding condition. While size alone does not mean functional effectiveness, during the annual cycle of a given species the two are closely correlated, so that a meaningful quantitative measure of reproductive readiness in most birds can be obtained by weighing the ovaries or testes.

Along with the size and physiological state of the reproductive organs, and the secondary sexual characters such as the breeding plumage, there is an annual, cyclic, fluctuation in the behavioral readiness for reproduction and the raising of the young. Many manifestations of readiness to breed reach a maximum in spring. These include interest in the opposite sex, searching for suitable nesting sites, establishment of a territory around the prospective nest from which other males are chased away, and, finally the characteristic song of the species. Migration is another part of this annual cycle. It is just at the time when most species of birds begin to change from the winter condition with small testes and ovaries, winter plumage, and little interest in reproductive activities, into the state and activities appropriate for springtime, that they leave their winter quarters

and fly whatever distance may be necessary to reach
the areas where they will mate and rear their young.

How Does a Bird Know when to Migrate?

However strongly motivated by its annual reproductive cycle it may be, a migratory bird still has the
problem of when to start each seasonal journey. One
might suppose that the immediate temperature,
weather, food supply, or other environmental conditions would influence directly a bird's migration
north or its preparation in other ways for mating.
It would seem equally reasonable to expect that as
the arctic summer draws to a close the migrants must
linger until colder nights or diminishing food supplies make their summer quarters unsuitable. Only
then would it seem sensible for them to begin the
return trip south away from the arctic tundras which
they reached by such strenuous exertions a few weeks
earlier. But most migratory birds actually start their
spring migrations in climates very different from
those prevailing on their breeding grounds. The many
species that winter in tropical areas must leave these
regions of relatively constant conditions at the appropriate season if they are to intercept a fleeting
northern summer. What tells a bird in a rain forest
in South America when it is time to start north so
as to reach the Canadian tundra just after the snow
has melted? The birds' performance is all the more
remarkable in view of the relatively accurate seasonal
timing which some species achieve. Though the
newspaper headlines about the swallows of Capi-
strano are exaggerated, careful records kept over several years show nevertheless that certain species arrive at their breeding areas within the same week of
each year, if not on the same day. The arrivals and

departures of most migratory birds are more closely correlated with calendar date than with the immediate weather conditions.

The start of fall migration is similarly unrelated to the immediate situation in which birds find themselves in the summer months, after their young have been raised to a state of self-sufficiency. Most often the weather is still mild, food supplies are abundant; everything would seem ideal for a leisurely recuperation from the strenuous effort of northward migration, nesting, and gathering food for a growing family. But far from sinking into such a period of relaxation, many species immediately start south again. For instance, shore birds appear in large numbers on the sand flats of coastal regions in temperate latitudes during August, while many local songbirds are still feeding nestlings. Many sandpipers and plovers are already halfway to their winter range in the tropics before temperatures have passed their summer maximum or food supplies have begun to dwindle.

The Length of Daylight

Among environmental factors that might provide the necessary timing signals for migrations one of the most obvious is the length of daylight. In temperate latitudes, where most scientists have lived and marshaled their thoughts, the time from sunrise to sunset is one of the most predictable events in a highly variable climate. About thirty years ago William Rowan, a biologist at the University of Alberta, in Edmonton, began to investigate the matter directly with appropriate experiments. Most Alberta birds are migratory, and the winter climate is sufficiently severe so that only a few species remain in residence. Rowan concentrated on two abundant migrants that nest in the Prairie Provinces, the slate-colored junco, a relative of the sparrow, and the common crow. Both species are relatively hardy and do not migrate to the tropics but winter in a wide zone roughly corresponding to the north-central United States. In spring both migrate north in large numbers not only to the latitude of Edmonton (54° N), but also as far north into Canada as low-scrub woodlands are to be found.

Although these birds normally migrate far to the south, they had little difficulty surviving temperatures as low as −52° F., provided sufficient food was readily available. Long summer days provide a real advantage to birds that migrate to the Far North, but short days and long nights make a northern winter dangerous for small birds, even those that are adapted to utilize seeds or other foods that do remain available. A major limitation on the survival of such birds under natural conditions during cold spells in winter is the length of daylight available for the gathering of food. Even in the relatively mild climate of Massachusetts sparrows eat enough seeds during the day to store in their bodies about 10 per cent of their body weight as fat. But during a cold night nearly all this fat is used up, even though the bird is asleep most of the time. Merely to maintain its body temperature requires consumption of this relatively large proportion of its own body weight. Very little added stress, whether lower temperatures at night, lack of roosts sheltered from the wind, or slightly less abundant food, can easily transform tolerable conditions of winter into absolutely fatal ones. These are among the problems that birds escape by migrating south.

Rowan added electric illumination to increase the

normal day length in aviaries housing his experimental birds. As a control he set up another aviary without any artificial illumination. Both experimental and control aviaries were unheated and unprotected from the winds and blizzards of the Canadian prairies. It had been widely argued that the rising temperatures in the spring were the essential factor that brought wild birds into breeding condition, and stimulated their northward migrations. Rowan suspected otherwise and wished to be sure that his birds were exposed to the full rigors of the Alberta winter. Beginning in early November, he lengthened the day for his experimental birds by five minutes each day until mid-December they were spending about fifteen hours in the light while the controls had only the nine-hour day nature provides at the latitude of Edmonton.

The results of Rowan's experiments were dramatically clear. His experimental birds responded to the increased day length with greatly enlarged ovary or testes. In mid-December the males were singing vigorously, as they would do normally in spring when defending a nesting territory. The control birds remained in the typical winter condition, both with respect to the size of their reproductive organs and their behavior. After the experimental juncos had been brought into what appeared to be full breeding condition in midwinter, both they and the controls were released, in order to see whether there would be any migration. The results were inconclusive. Some of the experimental birds and almost all of the controls remained near the aviaries and were retrapped; others disappeared, but there is no way of telling whether they migrated or failed to survive the severe winter conditions. If they migrated, there was no evidence concerning the direction of movement.

Rowan's Experiments with Crows

While Rowan's first experiments had shown that
he could bring juncos into breeding condition by
artificial increases in the length of daylight, they had
not produced any very convincing evidence about
migration itself. It was possible that his juncos migrated north in January after being thus experimentally brought into breeding condition. In an attempt
to test this possibility Rowan turned to crows, which
are more conspicuous birds. There was better reason
to hope that after the birds moved away from the
experimental aviaries, they might be seen by human
observers and reported. The same type of experiment
brought sixty-nine experimental crows into approximately normal breeding condition in early November, and fourteen controls were also available after
spending the autumn months in unlighted aviaries.
Both groups were released on November 9, when all
but a very few stragglers from the wild population
had long since migrated south to the general area of
Oklahoma. At the time of the release there was widespread newspaper and radio publicity about the
experiment. Everyone in Alberta was invited to
shoot crows and report their band numbers, and to
turn in the bodies for examination. The crows were
released just before the Thanksgiving weekend when
more hunters would have time for crow shooting.

Several crows were shot within a few miles of
Edmonton, too close to the point of release to indicate whether they might have migrated had they
been left undisturbed. Of the controls only four were
reported after flying any considerable distance; and
all those were found at points to the southeast,
in the normal direction of their fall migration. One

had flown about 200 miles on its way toward Oklahoma. Sixteen experimental birds were shot at substantial distances, and of these eight were south or southeast, while eight were north and northwest. The fact that equal numbers of experimental crows started to migrate north and south (that is, in fall and spring directions) makes the experiment difficult to interpret. To be sure, there was some possibility that the crows that flew south might not have been brought into full breeding condition by the experimental light schedule. One might conclude simply that the experimental birds were stimulated to fly for considerable distances regardless of direction, but it is certainly suggestive that none of the controls had moved northward while half the experimentals had. In addition to the eight crows definitely secured to the north and northwest, several others were seen as far as 300 miles north of Edmonton in the months of November and December. Crows are so rare in winter in the sparsely inhabited regions north of Edmonton along the Peace River and Lesser Slave Lake, that there is a very good probability that these birds seen but not secured were actually Rowan's experimental crows. The density of human population is so much greater south of Edmonton than to the north that a strong bias favored recoveries in the direction of the normal fall migration.

The most northerly of the eight crows actually shot was seen in February about 100 miles north of Edmonton by a trapper who had been living in the wilderness since early autumn, and had no idea that such an experiment was under way. The sight of a crow at that desolate season was so astonishing that he feared he might be succumbing to delusions from his solitary isolation. He therefore shot the crow to confirm his own sanity. Finding it banded, he saved the band and reported it later in the spring on his return to civilization.

At first it seemed that Rowan had solved completely the problem of seasonal timing in bird migrations. After some of his juncos had been brought prematurely into breeding condition in midwinter he shortened their day length with the result that they returned to the nonreproductive state in the early spring, just when they would normally have been beginning to breed. Thus it seemed that changes in day length could switch on or off the breeding activity of birds. But when Rowan's experiment was repeated with other species and at other seasons it did not always work. For instance, lengthening the birds' day in late summer or early autumn usually failed to produce enlargement of the ovaries or testes. There is considerable variation from one individual to another, and from species to species, but most birds respond to lengthening daylight only in winter and spring, and for some weeks or months after the season of active breeding they are in a refractory state.

Rowan chose for his experiments species that did not migrate south of temperate latitudes. In Oklahoma crows are subjected to increasing day length throughout the spring, and so are juncos in the latitudes where they spend the winter. But many birds migrate to tropical latitudes where they experience only slight changes in day length. Yet they begin their northward migration punctually every spring.

Still more difficult to explain solely on the basis of day length are the well-timed, northward departures of the many birds that migrate far south of the tropics every fall. They travel from very long days in midsummer—at the extreme from the continuous daylight north of the Arctic Circle—through tropical regions where the days are about twelve hours long, and then escape from late summer at the end of their journey into the early spring of the Southern Hemisphere. In October and November their days grow longer instead of shorter, but they do not begin to come into breeding condition until the months when spring comes to the Northern Hemisphere. At that time they are actually experiencing shorter and shorter days.

Although he was naturally elated at the success of his experiments, Rowan recognized this difficulty clearly, and, in 1931, he postulated that there were annual cycles in the reproductive activities of transequatorial migrants such that their refractory periods to increasing day length persisted until after they had started north and passed through the tropics. In these species something other than increasing day length in the strict sense must be the immediate stimulus to northward migration. Nevertheless, Rowan considered the size and functional state of the testes and ovaries to be the primary phenomenon, and he believed that migration resulted directly from high levels of the sex hormones known to be produced by ovaries and testes, and released into the blood, when these organs were at their maximum size. Some types of reproductive behavior can be stimulated under certain conditions by the sex hormones, and Rowan inferred that migration belonged in this category.

Experiments similar to Rowan's were carried out later by Albert Wolfson, who worked with closely related subspecies or geographical races of juncos that spend the winter in California. One of these is nonmigratory, but the others fly north every spring

to nest along the Pacific coast of Canada and southern Alaska. Wolfson found, as had Rowan, that
juncos came into breeding condition prematurely
when kept on long days through the winter, whether
they belonged to a migratory or nonmigratory race.
When these birds were released, the migratory
juncos disappeared, and one was recovered 200 miles
to the north. The nonmigratory juncos seemed to
remain in the general vicinity. This experiment
showed that even though already in full breeding
condition birds may nevertheless carry out a northward migration, provided that they have the heredi-
tary make-up of a migratory species.

The Influence of the Pituitary Gland

The notion that there is a truly simple, direct
cause-and-effect relationship between the level of sex
hormones and migration was later ruled out by experiments in which castrated birds migrated almost
normally. Other experiments suggested strongly that
migratory behavior is not a direct result of sex hormones, but that both growth of the ovaries and
testes, and migration are regulated by hormones
from the pituitary gland. This important gland is at
the base of the brain, and the hormones it secretes
into the blood stream regulate not only the state of
activity of ovaries and testes, but the states of many
other important organs of the body. Furthermore,
the secretory activity of the pituitary gland is itself
closely interrelated to the functioning of the brain,
particularly a closely adjacent portion of the brain
called the hypothalamus, which is especially concerned with reproductive behavior. It is far beyond
the scope of this book to discuss endocrinology, but
appropriate experiments have shown that light stim-

ulates the pituitary gland to release hormones that, in turn, promote growth of the testes and ovary. This effect of light is ordinarily through the visual sense of the bird, but even when birds have been blinded light penetrates directly to the pituitary gland or adjacent parts of the brain and produces the same type of stimulation. The lengthening days of spring exerted a powerful and basically important effect on the whole complex system of brain, pituitary gland, reproductive organs, and the appropriately related behavior of the bird, including long-distance migration.

Since Rowan's pioneer experiments other biologists have accumulated evidence supporting his assumption that some months after the end of the previous breeding season and fall migration birds undergo a gradual but spontaneous process of increasing readiness for reproduction and spring migration. For instance, Wolfson found that if birds were artificially exposed continuously to days as short as those of mid-December, they would eventually return to breeding condition, but at a much slower rate than normal. Evidently the increasing day length of the natural spring accelerates a process of annual change that is already underway within a bird's body, and brings it into breeding condition in March, for example, rather than in June. Rowan's original dramatic experiments revealed not a single, unique cause of reproductive activities and migration, but one crucial timing factor responsible for keeping birds synchronized with the seasonal climatic cycles. It seems likely, though far from certain, that this type of intrinsic cycle of slow return from the nonbreeding, winter condition to the springtime state of endocrine and behavioral activity progresses at a slightly different rate in various species of birds.

Probably those that spend the winter in the tropics, or even in the long days of temperate latitudes in the Southern Hemisphere, have their annual cycles timed more by these spontaneous, internal changes than by the increasing day length characteristic of winter and early spring in north temperate climates.

Stored Fat as Fuel for Migratory Flights

Shortly before birds begin a major migration they amass considerable quantities of fat, either just beneath the skin, or in concentrated fat deposits within the abdominal cavity. This accumulation is so characteristic and conspicuous that it serves better than any other known criterion to distinguish birds that are just about to migrate from others that will not do so in the immediate future. Deposition of fat is part of the whole set of bodily changes produced in migrants by artificially lengthened days during the winter. The energy for flight, like that for any other activity carried out by a living organism, is derived basically from the oxidation of foodstuffs. Almost all the foods consumed by animals of any sort are carbohydrates (sugars and starches), proteins, or fats. Fats supply nearly twice as much energy per unit weight as carbohydrates or proteins, and it is thus an obvious economy for a flying animal to rely on fats as stored food. The stored fat of birds, particularly small songbirds, is readily available, and it is transferred quickly from the cells where it is stored into the blood stream for oxidation in muscle cells, or elsewhere as needed. On a long flight appreciable quantities of the stored fat are thus consumed, just as on a very cold winter night a small bird may use up most of its bodily fat merely to keep warm until morning. In Wolfson's experiments the nonmigratory race of juncos did not lay down additional fat when experimentally exposed to lengthened days during the winter, while the migratory juncos bulged with stored fat.

Other measurements of the amount of fat in the bodies of birds at various stages of their migrations illustrate emphatically the magnitude and importance of these compact fuel reserves. White-throated sparrows, analyzed just before their spring migration, had an average of 17 per cent of their body weight in fat, while the same species just after completing its migration from the southern United States to Canada had only 7 per cent. Studies of other white-throated sparrows that were killed in collision with a television tower in Florida on their fall migration showed a fat content of only 6 per cent. These sparrows were close to the end of their southward migration and presumably had used up most of their fuel reserves. But several species of warblers killed the same night at the same tower averaged 30 per cent fat; these birds would have continued on south for several hundred miles to the Caribbean islands or the mainland of South America. Their fat reserves were only slightly depleted at the time of their accidental deaths in Florida. Hummingbirds, the smallest of migratory birds, accumulate relatively enormous quantities of fat in their tiny bodies. Analyses in June, during the nesting season, showed only 11 to 15 per cent fat, while in late summer hummingbirds from the same populations contained 41 to 46 per cent fat. These exact figures should not be taken too literally, as such measurements also reflect minor differences in the available food supply, the time of day when a bird was taken, and many other factors.

To what degree are these bodily fat reserves adequate for the actual energy requirements of major migrations? To answer this question we would have to know the rate of energy consumption of the birds during actual flight. Since virtually all the energy production of higher animals depends upon the complete oxidation of foods to carbon dioxide and water, the most accurate and convenient method for finding rates of energy production, usually called metabolic rates, is to measure the rate at which oxygen is taken into the lungs. Despite the difference in energy released per gram of food, the energy obtained by combination of any food with a given amount of oxygen is nearly the same. Hence, the rate of oxygen consumption can be converted directly into metabolic rate, or rate of energy made available for all the bodily activities. Some oxygen is needed, of course, for the basic maintenance of bodily function, and this amount approximates the resting oxygen consumption. But in strenuous exercise, such as the flight of birds, it is a relatively minor component. Unfortunately, however, we have almost no information about the metabolic rates of flying birds, a somewhat unexpected gap in our knowledge in view of the considerable research effort that has been devoted to other aspects of bird migration, and to the biology of birds in general. The difficulty is that none of the standard methods is easily applied to birds in free flight. Most birds will not fly in any normal fashion in chambers small enough for their metabolism to produce a reliably measurable change in the level of oxygen. Nor do they fly at all with face masks and rubber tubes attached. Until some new effort or technical ingenuity is applied to this problem, we must fall back on indirect evidence.

The closest available data come from oxygen-con-

sumption measurements carried out with hummingbirds hovering in glass jars containing a little more
than one gallon of air. An actively hovering hummingbird consumes oxygen at a rate that would
"burn" its fat reserves at a rate of about 0.13 gram
per hour. Hummingbirds closely related to the species used in these measurements migrate across
the Gulf of Mexico, apparently covering at least 500
miles in one night's flight. These birds have been
observed to fly at about fifty miles per hour, so that
ten hours' flying could take them across the Gulf. If
we make the assumption that the metabolic cost of
migratory flight at fifty miles per hour is the same as
that of hovering flight, we can estimate that this
flight across the Gulf of Mexico would consume 1.3
grams of fat. This assumption may be in error in
either direction; forward flight probably allows some
lift from the motion of the wings through the air,
but this must be at least partly offset by the air re-
sistance to be overcome in traveling at fifty miles
per hour. Hummingbirds captured just before starting this migration sometimes contain as much as
two grams of fat; hence, there seems to be a small
margin of safety. These two grams constitute over 40
per cent of the body weight, however, and the hummingbirds appear to be stretching their fuel reserve
to the limit in order to achieve this particular overwater flight.

Migration Restlessness

Most laboratory studies necessarily require the
confinement of migratory birds in small cages. Since
most birds depend heavily on vision to guide their
daily activities, it is not unexpected that in total
darkness they should remain almost stationary on

their perches. The same is true under most conditions if a steady, dim light is provided during the night to approximate the faint illumination usually available under natural conditions from the moon or stars. But migratory birds sometimes show considerable activity at night in the seasons when, if they were free, they would be migrating. This activity usually takes the form of rapid flitting back and forth in the cage, provided that several convenient perches are available. Or the more nervous species may actually flutter against the walls of the cage at such times, even though at other seasons they are sufficiently adapted to the conditions of captivity to avoid such apparently pointless behavior. This characteristic jittering about is called *migration restlessness*. Experimental increase in day length during the winter months induces migration restlessness along with the other changes that have been described, such as accumulation of fat reserves. It constitutes one more sign that a bird is physiologically in a state of readiness for migration. Migration restlessness has been put to good use in experiments designed to analyze the methods of navigation used by migrants, as will be described in later chapters.

Chapter 4

BIRD NAVIGATION

How does a bird set and hold an appropriate course on a long migratory flight? Few questions have fascinated so many thoughtful people, but only a partial answer can yet be offered. Some aspect of a bird's environment must convey information about the proper direction for its migration. Something in the outside world must be related to this direction, generally north in spring and south in fall, although the precise direction varies from species to species and from one to another part of many migration routes. If we try to imagine ourselves in the bird's position, we cannot think how we could find our way without artificial instruments. It is not enough for the environment to provide some cue to the appropriate direction of migration. Such a cue would be useless unless the bird were equipped with sense organs capable of reacting to it. But in considering the sense organs of birds in relation to their problems of long-distance navigation, it is essential to bear in mind that they are vertebrate animals, with nervous systems built according to the same general plan as those of mammals, including ourselves.

Subtle Sensory Factors

This basic similarity of all vertebrate animals does not preclude minor, but critically important, differences in modes of orientation, comparable for instance to those that underlie the orientation of bats. Since bats are mammals, they have nervous systems even more similar to our own than those of birds. Yet a relatively small change in the frequency range of their hearing and sound production enables them to operate a highly refined sonar system, which until recently remained unrecognized merely because it operates one to three octaves above the range of human hearing. (The acoustic orientation of bats and blind men is described more fully in *Echoes of Bats and Men*, Doubleday Science Study Series, 1959.) One of the standard vertebrate sense organs might be slightly modified in birds to guide them during long migratory flights. The difficulty is to identify such a sensory specialization adequate to the demands that migration would place upon it.

Another sort of explanation requires no special sensitivity on the bird's part not shared, at least to some degree, by other vertebrates, including ourselves. This type of explanation seeks to identify some subtle aspect of the bird's environment—some sight or sound or smell—which a human observer might find quite detectable if he were alert to its importance. Honeybees guide themselves by such an environmental cue, the polarized light from the blue sky. They can, in certain circumstances, determine directions by the way in which the pattern of polarized light is related to the position of the sun. (See Karl von Frisch, *Bees, Their Vision, Chemical*

Senses, and Language, Cornell University Press, 1950.) Ordinarily we cannot tell polarized from unpolarized light, much less judge the plane of polarization. But there is an obscure phenomenon of human visual physiology known as "Haidinger's brushes," through which people with normal vision learn to distinguish the plane of polarization of light, though only under favorable conditions. The simplest way to demonstrate Haidinger's brushes is to mount a sheet of polaroid immediately behind a uniformly illuminated translucent surface, such as a piece of ground glass, and hold the polaroid in a device that allows it to be rotated about once per sec-

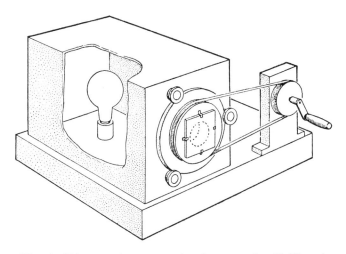

Fig. 10. Diagram of apparatus for demonstrating Haidinger's brushes. A bright light source is enclosed in a light-tight box. The crank rotates a piece of polaroid through which the light from the source is transmitted to the observer. To avoid confusion due to seeing the actual rotating parts, a piece of ground glass (not shown here) should be placed just in front of the apparatus. The observer will then see a uniformly bright spot of light. The demonstration is most effective when a blue filter (not shown in the diagram) is placed between the lamp and the polaroid.

ond. It is further necessary that the light source be shielded, as shown in Fig. 10, so that the ground glass is illuminated only from behind, by the polarized light. If the polaroid is stationary one sees a uniform field, but as it rotates most people see an ill-defined rectangular spot, or sometimes a sort of Maltese cross rotating at the same rate. The effect is more pronounced if blue light is used.

On a clear day the blue light of the sky is partly polarized, the maximum percentage of polarization being present at roughly ninety degrees from the sun (for example, close to the zenith when the sun has just risen or is about to set). Some people can see Haidinger's brushes by looking at the blue sky, although most find it difficult, and the vague image fades rapidly when one concentrates attention on it. A trick that sometimes helps is to rotate the head as nearly as possible around the axis of one eye (an entertaining way to exercise the neck muscles). This movement has much the same effect as rotating the polaroid sheet in the simple demonstration described. Just how plane polarized light stimulates the human eye to perceive Haidinger's brushes is a matter of controversy. But it is clear that if this were an important accomplishment, people could learn to do it with some reliability. It seems unlikely that polarized light could guide birds during migration, since direct tests have failed to show any sensitivity of birds to polarization of light. But might there be other, equally subtle factors within the sensory horizons of migratory birds which give them directional information?

The sense organs and nervous systems of birds have been studied carefully in search of specializations that might be related to their ability to navigate over long distances. The sense of smell is very poorly developed, as exemplified by the relish with which great horned owls prey and feed on skunks, despite the victim's chemical defenses. Most birds have excellent hearing, and they are equipped to feel extremely faint vibrations of the ground or any object on which they are perched. As in all vertebrate animals, the inner ear is a series of interconnected, fluid-filled tubes or chambers containing specialized sense organs that respond to accelerations. One of these is the cochlea, which is sensitive to sound waves. This inner-ear labyrinth, as it is called, consists basically of clusters of mechanically sensitive cells. All the cells in each group are linked by hairlike projections to a mass of secreted material floating in the fluid content of the labyrinth. Four of these groups of hair cells are specialized for response to linear accelerations or decelerations; the other six are found at points along tubes called semicircular canals. The fluid in these loops tends to rotate relative to the walls whenever the bird's head rotates around the axis of the particular canal. Since the six canals lie in different planes within the head, there is ample opportunity for the stimulation of sensory nerves during every type of rotating maneuver that a flying bird can execute.

We, too, have an inner-ear labyrinth, but the sensations originating in it seldom reach our consciousness except under the special conditions such as when, among other disturbing symptoms, we experience sensations of movement that do not accord with what we see. It would not be surprising if the inner-ear labyrinth of birds were specialized in some way for a life on the wing, but to date no convincing evidence has been accumulated to show that it differs in any important way from our own. Perhaps

this state of affairs results from insufficiently critical studies of the inner ear of birds.

Bird Vision

The eyes of birds are enormous relative to the rest
of the head. At least in small species of birds the
eyeballs are almost as large as the brain. The optic
nerves leading from the eyes to the brain are also
large, and the parts of the brain to which these
nerves ramify are equally well developed. Since a
brain contains innumerable, minute, and vastly complicated networks of nerve cells and their intercon-
nections, it is impossible to establish the exact limits
of the visual areas. But these are certainly far larger
in birds than the brain areas devoted to hearing, or
any of the other senses. Although the large size of
birds' visual machinery suggests an overriding importance of vision, one must be cautious about applying this anatomical fact directly to the problems
of bird navigation. It is quite conceivable that the
enlarged and refined visual system justifies its exist-
ence in other aspects of the bird's life, such as the
location of food, or the avoidance of predators.

It is surprising to find one prominent structure in
the bird eye which appears to *subtract* rather than
add to its visual capabilities. The structure in question is called the pecten, and a typical example is
shown in Fig. 11. It is almost entirely an avian pecu-
liarity, although many of the reptiles possess a relatively smaller structure of a similar nature. It consists
of a large, blood-filled shelf projecting inward toward the center of the eyeball. It is attached directly
over the entrance of the optic nerve. In most birds
the base of the pecten occupies a strip of what would
otherwise be functional retina. Where a fish or a

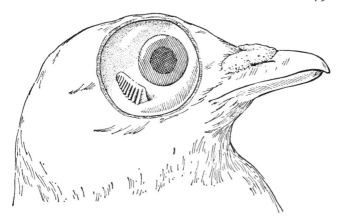

Fig. 11. A bird's head and eye with feathers, skin, and part of eyeball cut away to show the relative size of the pecten projecting from the lower half of the retina. Note also how large a fraction of the entire head is occupied by the two eyes.

mammal would have thousands of light-sensitive cells, and nerve cells, the bird has a huge blind spot. The pecten usually has a complex shape with many folds resembling the bellows of an accordion. It also contains heavy deposits of pigment, and sometimes is brilliantly colored though often virtually black. In some birds it covers a narrow band extending 60 to 90 degrees obliquely upward across the inside surface of the eyeball. The pecten itself is entirely devoid of nerve cells, and cannot be stimulated by light. Why should the bird eye, which is so large and so obviously specialized for excellent vision, have the largest-known blind spot? Biologists have not yet, in my opinion, provided a satisfactory answer to this perplexing question.

Granted then that birds have excellent vision (except for the large blind spots due to the two pectens), what is there for a migratory bird to see that could tell it the direction appropriate for its migra-

tory flight? Some migrations proceed along coastlines, and river valleys, or parallel to ranges of
mountains. Perhaps these or other visible features
of the earth's surface supply natural signposts to
mark migration routes. But much migration, particularly of smaller birds, occurs at night, and in many
other instances the birds migrate over the open
ocean, which is, of course, devoid of any obvious
topographical features. Even over the ocean, however, there may be patterns of waves or fairly stable
cloud formations related to consistent trade winds
that conceivably might be helpful. And it always remains a possibility that there is some other visible
sign, comparable to Haidinger's brushes, available
from the sea or sky, which human observers have
not yet noticed. Nevertheless, the fact that so much
migration occurs at night makes such theories of topographical guidelines extremely difficult to apply to
birds in general.

Do celestial bodies, the sun, or stars, serve as bea-
cons to guide migratory birds? This idea was suggested long ago, in the first century of scientific biology, but it was never taken very seriously because
all the celestial bodies except the North Star and a
few of its close neighbors change their apparent positions in the sky as a direct consequence of the earth's
rotation. It certainly seems farfetched, at first
thought, to postulate that birds might be sufficiently
familiar with the stars to pick out the North Star
among all others as a fixed reference point. It seems
equally rash to speculate that they might correct for
the motions of sun and stars during the night and
day, as a mariner does with the aid of his chronometer and astronomical tables. Even after the motions
of the heavenly bodies were adequately understood,
practical celestial navigation became possible only

when chronometers were perfected to the point that they would keep extremely accurate time over long periods. The notion that birds might possess in their minute brains the equivalent of sextant, chronometer, navigational tables, and the knowledge which human seamen study diligently for months to master —all such postulates seemed preposterous. Nevertheless, in the next chapter we shall see that the capacities of birds in this respect were seriously underestimated.

Orienting Thermal Radiations?

Although we shall return to types of orientation that depend upon vision, it is of interest first to consider several other theories that have been advanced to account for bird navigation. One often helpful way to approach a scientific problem is to relax and take the broadest possible view, starting from the most rudimentary considerations. The summer homes of migratory birds are usually in cool regions compared to their winter ranges. Now it is well known to physicists that all objects emit electromagnetic radiation, the intensity and wave length of which depend on the temperature of the source. Cool objects radiate longer wave lengths at lower intensities. Thus the arctic regions, toward which birds migrate in spring, radiate less energetically than the tropics. Let us suppose that birds possess adequate sense organs to detect these types of radiant energy, and that these sense organs allow them to perceive the "warmer" radiations from the south. How delightfully simple an explanation of bird migration this would be!

When one considers what sort of thermal radiation actually would flow from a warm climatic region to a cooler one, several serious difficulties become apparent. First of all, the earth is curved, and radiation tends to travel in straight lines. It is thus difficult to suppose that such radiation could reach the bird from more than a few tens of miles, at the very most. Furthermore, the temperatures of air or land or sea are low enough that the infrared or "heat radiation" has wave lengths on the order of ten micra. (One micron equals 10^{-6} or 1/1,000,000 meter; visible light has wave lengths ranging from about 0.4 to 0.75 micron.) But several studies of the visual sensitivity of birds to various wave lengths of light show that their spectral limits agree very closely with our own. Hence, they must be quite insensitive to the wave lengths required by this theory. A further difficulty is that climatic temperature variations do not become pronounced until we look at them over very large distances north and south. Lakes, deserts, wooded areas, or even cities produce drastic distortions over distances of a few miles. A bird just north of Lake Superior in the spring or summer would receive "warmer" radiations from the north than from the air above the cold waters of the lake. This theory thus fails miserably on both counts—the physical signal from the environment would be hopelessly ambiguous, and there is no evidence that the birds' nervous systems could respond to it at all.

Terrestrial Magnetism?

Another superficially attractive theory, already mentioned, is that birds possess some equivalent of a magnetic compass. This hypothesis has taken several forms, in addition to the obvious one that some part of the bird's nervous system, like a compass needle, is stimulated by the relatively weak magnetic field of the earth. Since the earth's magnetic field is not in general parallel to the ground, a compass needle dips downward at one end if it is free to tip vertically as well as to turn horizontally. There are variations over the earth's surface in the angle of this dip or declination, and some theorists have found it easier to postulate a sensitivity to declination rather than to the direction of the magnetic field. A different approach is based on the fact that any electric conductor moving through a magnetic field is subjected to an electromotive force that tends to move charged particles within the conductor. This, of course, is the principle upon which an electric generator operates. It is possible to calculate the magnitude of the electrical voltage that would be set up within a bird's body as a result of a given velocity of flight through the earth's magnetic field. For a bird flying forty miles per hour this voltage would be about 10^{-5} volt per centimeter.

While the electric currents induced by flight through the earth's magnetic field are extremely small, it is dangerous to state dogmatically that no part of any bird's nervous system could possibly respond to them. It was easier to be categorical in denying such a possibility a few years ago, before it was discovered that certain fishes generate weak electric currents in the water around them and use these currents to orient themselves with respect to objects at close range. They do this by sensing differences in the electric field that results from the differing electrical conductivity of objects in the surrounding water. Direct experiments with these fish have shown that in some circumstances they respond to voltage gradients as low as 10^{-7} volt/cm. This sensitivity arises from specialized cells of the lateral line, a sensory system also used by fish to detect water currents. No comparable sense organ or electrical sensitivity is known in birds, however; the extreme sensitivity of these specialized fishes can only serve to warn us against dogmatic incredulity in evaluating these theories of bird navigation.

If any of the theories of birds' sensitivity to the earth's magnetic field are correct, one should be able to stimulate birds by means of artificial magnetic fields. Or, if they showed no spontaneous response to a magnetic field, one would expect that they could be trained to respond, provided they possessed the necessary magnetic sense organs. Several such experiments have been attempted with a wide range of magnetic fields, from approximately the magnitude of the earth's field to extremely high magnetic intensities. Both constant and varying fields were used. The latter would induce currents in the bird much as would flight through the earth's magnetic field. Other experiments in which magnets were attached to the birds themselves will be described in Chapter 5. The outcome of all these experiments has been completely negative. While no one can be absolutely certain that another experiment of this sort, with some different type of bird or under different experimental conditions, might not produce evidence of sensitivity to magnetic fields, the weight of evidence is strongly against such a supposition.

One apparent exception to this statement should be mentioned. If the magnetic field is many hundreds of times stronger than that of the earth, induced currents can reach proportions sufficient to stimulate eyes and other parts of the nervous system directly. Men close to very large magnets sometimes experience visual sensations that seem to result from such induced currents. Similar effects are probably produced when birds fly through a concentrated radar beam, but only at intensities of electromagnetic radiation that are so far above any present in the normal surroundings of migrating birds that they would have no significance whatever in the behavior of birds under natural conditions.

Orientation Based on the Earth's Rotation?

Another equally ingenious, but more complex, theory was advanced by a Swedish physicist, Gustaf Ising, shortly after the Second World War. He suggested that birds might be sensitive to the purely mechanical effects of the earth's rotation, that they could not only detect such effects, but distinguish their variation in strength with geographical latitude. This type of theory can take several superficially different forms, although on close examination they all reduce to much the same thing. One of the easiest to understand is a hypothetical variation in a bird's apparent weight as it flies east or west, because its flight speed is added to, or subtracted from, the speed of rotation of the whole surface of the earth, including its atmosphere. In thinking about Ising's and other similar theories, it is very helpful to have available a small globe. If the sort made specially for studying geography is not available, a useful substitute can readily be fashioned from a grapefruit impaled on a knitting needle.

As newspaper readers know in these days of artificial satellites, any object that maintains a horizontal velocity of approximately 18,000 miles per hour goes into an orbit around the earth, rather than falling to the ground. An old-fashioned way of describing this phenomenon was to say that anything moving along a curved path tends to enlarge its circle of rotation because of a "centrifugal force." On this basis one would say that an artificial satellite defies gravity because the centrifugal force resulting from its high speed balances the gravitational attraction of the earth. Physicists now prefer to look at the matter somewhat differently, by emphasizing the fact that any moving object can be deflected from its path only by an acceleration operating in some other direction than the straight line along which it is traveling at the time. If an object is to move at a constant speed around a circular path, it must be subjected continuously to an acceleration directed inward toward the center of the circle of rotation. This *centripetal* acceleration must always operate at right angles to the direction of the object's motion. The modern physicist regards the satellite in a circular orbit as tending to continue its rapid motion in a straight line, except that the earth's gravitational attraction provides just enough centripetal acceleration to hold it at a constant distance from the center of the earth.

Of course, no bird flies at anything approaching orbital velocity. But whenever a bird flies east it adds its speed of flight to the speed at which the surface of the earth rotates around the earth's axis. Conversely, when flying west, the bird's air speed must be subtracted from the rotation of the earth in order to find the true speed of rotation of the bird around the axis of our planet. The force with which our feet tend to press against the ground or the force exerted by a bird's wings against the air would be very slightly greater if the earth ceased its rotation. This is the case because a very small fraction of the gravitational attraction is necessary to deflect us inward toward the center of the earth, instead of our continuing in a perfectly straight path out into space. Sixty miles per hour is an approximate maximum air speed for a fast-flying bird. This is about 1/300th of the orbital speed of 18,000 miles per hour at which an object near the earth's surface has an apparent weight of zero, because the gravitational attraction is only just sufficient to curve its path inward and hold it in motion parallel to the ground. At the equator, if all other factors are exactly the same, the apparent weight of a bird that flies east at sixty miles per hour should decrease by about 1/2700th (assuming that the surface of the earth rotates at 1000 miles per hour). In directions other than east or west the changes will be smaller, with none at all during flight directly north or south. If these changes could be sensed by the flying bird, it might theoretically be able to circle and judge at once the points of the compass.

How might a bird flying at high speed be able to estimate its own weight? Obviously, when other factors are equal, its weight might be felt as a mechanical pressure of the air against the wings. This pressure, however, varies enormously with velocity of flight, and also, during flapping flight, with the cycling movements of the wings. Furthermore, any slight turbulence in the air would produce upward or downward accelerations of equal or, more likely, much greater magnitude. Since the earth's rotation cannot be varied experimentally (and perhaps this is just as well!), it is difficult to evaluate this hypothesis by direct test. Furthermore, we have so far considered only the maximum difference at the equator between flight at sixty miles per hour east and west. Many birds, of course, fly much more slowly, and most migration occurs at higher latitudes.

At the North or South Pole there is no rotation of the earth's surface or of the atmosphere directly above it. Therefore the bird's apparent weight will not be affected by its direction of speed of flight. On the other hand, the earth is rotating beneath a flying bird. Hence, if it flies in a straight line for any distance, the earth will meanwhile have moved on. Looking down at the snowy surface, the bird might notice that it was tracing a curved course rather than a straight line. Provided the flight is truly straight (relative to the axis of the earth), the more slowly the bird flies, the sharper the curvature. Suppose for example that a hummingbird flew at five miles per hour in a perfectly straight line for twenty-four hours over the vicinity of the North Pole. During this time the earth will have made a complete revolution, and the bird, though flying straight, will appear to have traveled in an almost complete circle over the ice-pack. (The circle would not be perfect, because in flying 120 miles the bird would have departed from the exact pole.)

Birds do not migrate across either the North or the South Pole, and most migration occurs at intermediate latitudes, where the effects of the earth's rotation are a combination of the two extremes outlined above—the rotation of the earth underneath the bird, so that it describes a curved path over the surface when actually flying straight, and the change in its apparent weight due to its direction of flight. The first of these two effects, called the Coriolis effect, is important in the motions of many real objects over the surface of the earth, including air masses and icebergs. Adequate consideration of the Coriolis effect on wind direction is beyond the scope of this book, but an explanation can be found in any elementary textbook of meteorology. The case of the iceberg is perhaps closer to that of the migrating bird. In certain parts of the oceans currents flow almost directly south (for example, the Labrador current). Fragments of ice breaking off from glaciers in Greenland are carried south by this current, but over a period of days they tend to turn clockwise, to the right, because of the Coriolis effect. Of course, many other factors such as winds and variations in the direction of the current due to bottom topography can also affect the movement of icebergs. But the Coriolis effect is sufficiently important and consistent that it is routinely taken into account by the Coast Guard in predicting the probable course of drifting icebergs.

If winds and icebergs can be deflected by the Coriolis effect, it is rash to exclude categorically the possibility that the effect might be perceived by flying birds. Just as in the change of apparent weight, however, the magnitude of the effect is very small, and it is difficult to see how it could be distinguished from mechanical forces acting on the bird's body. Any slight wind, any difference in lift or drag of the two wings would deflect the bird in a way that would impede greatly, if not fatally, any determination of the slight curvature of its path over the surface of the earth. Thus the earth's rotation theories are not taken at all seriously by biologists concerned with the explanation of bird navigation. But we are on less solid ground in rejecting such theories than in the case of the infrared or magnetic hypotheses. The fact is that no one has managed yet to devise a workable experiment by which these theories based on the earth's rotation can be put to direct test. The magnitude of the apparent weight change due to the earth's rotation will of course be proportional to a bird's velocity, and a stationary bird, or one moving very slowly, should have much greater difficulty than our hypothetical example flying at sixty miles per hour. Yet there is no evidence at all that high-speed flight, or even flight at all, is necessary for accurate orientation, as will be explained in the next chapter.

Chapter 5

HOMING EXPERIMENTS

For centuries homing pigeons have been bred, selected, and trained to carry messages in small capsules attached to their legs. Until the development of the electric telegraph, carrier pigeons provided the most rapid form of long-distance communication known to man, and even now pigeon racing is a popular sport. Furthermore, pigeons have continued to be useful for emergency and military communications.

The birds routinely are raised in relatively small rooms or sheds called "lofts," a given loft being the permanent home of its pigeons from birth to death or retirement from racing to become breeding stock. Only a swinging door stands between the pigeons and the outdoors, and the young birds are allowed to fly locally as soon as they are able. When the birds are three to six months old, systematic training begins. The trainer, carrying the birds in closed boxes or baskets, takes them some distance away and releases them for flight home to the loft. Some fail to return from these early exercises, but those that do succeed in homing become increasingly proficient. After perhaps a dozen flights from five to ten miles, the distance of the release point is increased rather rapidly. Many pigeons make successful flights from a hundred miles or more in the first season of their lives. Young and inexperienced birds often are released with older ones that have previously flown the course.

A firm attachment to a particular loft is essential for successful homing, and, once the attachment has been established, it is extraordinarily difficult to train the pigeons to home to another spot. Usually, when a new colony is to be started, breeding stock is taken from other lofts but is kept in permanent confinement, and only the progeny are allowed their freedom for subsequent training in the art of homing. Adequately trained pigeons will home at any season whether or not they are in breeding condition, have eggs to incubate, or young to feed. The most highly trained birds are quite capable of homing from five hundred miles or more in a single day, and there are records of returns from one thousand to fifteen hundred miles in two or three days.

Pigeon Racing

The highly developed competitive sport of pigeon racing shares some of the attributes of horse racing, on the one hand, and almost equally the spirit characteristic of 4-H clubs on the other, where enthusiasts compete more or less amicably against each other to see who can raise the best livestock. To raise pigeons that will win races requires the skill of a selective breeder, thoughtful care of the birds' health and welfare, and an undefined knack in the training for homing itself. It is common practice in the United States and Western Europe for a club to engage in competitions between lofts scattered throughout a large city and its suburbs. Since the various lofts are at slightly different distances from the release point for a given race, which may be fifty to five hundred miles away, the distance from the release point to each loft is divided by the elapsed time required by each bird, and the competition is based on this velocity of homing rather than the absolute time. Even without tail winds, average speeds of forty-five miles per hour over several hundred miles are not uncommon.

Each racing-pigeon loft is equipped with a timing mechanism. A temporary band is attached to the bird at the start of the race and removed as the pigeon enters the swinging door of its home loft. The band is inserted immediately through a little window of the sealed timing mechanism and clipped to a moving tape inside. On the evening following a race when the competitors have assembled at the clubhouse, judges open the timing devices, compute the birds' speeds, and award the prizes. Side bets are popular. Quite often a bird from a tiny loft on a tenement roof carries away the honors rather than the $500 thoroughbred from the large and expensively equipped loft of some wealthy member.

The homing ability is not limited to domestic pigeons. Many species of wild birds are able to return to their nests after being transported in closed boxes to a considerable distance. But unlike the year-round homing of the domestic pigeon, such homing seems to occur primarily when the birds are nesting. Since the male and female of most species participate in the care of the young, a homing experiment does not disturb unduly the domestic tranquillity of nesting wild birds. While one of a pair is in confinement or making its way home from the release point, the mate continues (at least for several days) to look after the eggs or the young.

In all careful homing experiments the bird is

TABLE I

TYPICAL HOMING PERFORMANCES REGISTERED BY VARIOUS SPECIES OF WILD BIRDS

Species	Number of birds tested	Distance transported (miles)	Per cent returning	Typical speed (miles per day)
Leach's petrel	61	135–470	67%	30
Manx shearwater	42	265–415	90%	200
Laysan albatross	11	1665–4120	82%	200
Gannet	18	213	63%	100
Herring gull	109	214–872	90%	60
Common tern	44	228–404	43%	125
Swallow	21	240–310	52%	150
Starling	68	200–440	46%	25

transported in a ventilated but opaque box from which it cannot see the surroundings at any time on the journey. The trip is made as quickly as possible—by boat, auto, train, or plane—so that the bird will be in good condition on release. Usually the bird is marked, with a colored band or a harmless dab of color on the feathers, to make spotting easy and recapture unnecessary when the bird returns to the vicinity of its nest. In most homing experiments it has been possible only to note the time of release and the time when the bird returns to its nest, if it returns at all.

The Phenomenal Manx Shearwater

Some of the most spectacular homing performances registered for wild birds are those of the Manx shearwater, which spends almost all its life at sea and is thereby a truly pelagic bird. Shearwaters feed on small fishes and crustaceans, which they catch by dipping an inch or so below the surface with their bills. They can rest on the surface of the ocean, but they are much more likely to be seen from ships as they fly in and out of the troughs between waves, seldom rising more than a few feet above crests. They build their nests in burrows in the ground, usually on offshore islands. They come ashore only at night, and only in the nesting season. Manx shearwaters from a small island called Skokholm, a few miles off the southwest corner of Wales, have returned from several release points along the coasts of England and France. These homing flights after artificial displacements often commonly covered 200 miles in twenty-four hours.

Two Manx shearwaters from this island off Wales were released at Venice, Italy, at the head of the

Adriatic Sea. Although one was never seen again, the other was back at its nest after only fourteen days. To estimate its speed, one must guess whether it flew a direct airline course over the continent of Europe, or remained over the ocean (as one might expect) and flew south of the tip of Italy, through the Straits of Gibraltar, around Portugal, and thence north to Wales. The airline distance home across the Alps and France is about 900 miles, while the shortest route over salt water is 3700 miles. Normally shearwaters are never observed flying over the land except after violent storms that have blown them inland; the overland route seems most unlikely. Assuming that the bird did not depart from the ocean on its fourteen-day journey, it must have averaged about 265 miles per day.

Two other shearwaters were transported by airplane from Wales to Boston, Massachusetts. Again one did not return, or was not observed at its nest. But the other reappeared only twelve and a half days after its release, clear across the Atlantic. Its minimum average speed was 244 miles per day. To be sure, the prevailing winds are from the west to east over the North Atlantic, and the favorable winds may well have assisted the bird. These two homing flights, from Venice and from Boston, are all the more impressive, because both release points are, in all probability, outside the range of the species. The shearwaters of the Mediterranean are held by some specialists to be a slightly different race, and while Manx shearwaters have occasionally been recorded off the east coast of the United States, they are so rare that it is extremely unlikely that this particular bird has ever visited the western shores of the Atlantic.

Still more impressive homing flights than those of

Manx shearwaters have been made by albatrosses from the Midway Islands, in the Pacific Ocean west of the Hawaiian Islands. Eighteen of these large pelagic birds were taken from active nests and transported by airplane to the distant points in the Pacific shown in Fig. 12, among them Oahu, one of

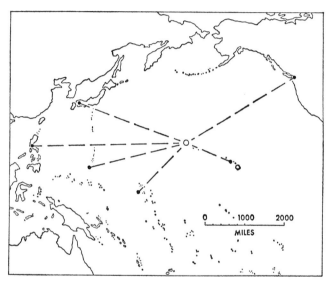

Fig. 12. *Homing flights of albatross. Birds captured at their nests on Midway Island (open circle) were transported by airplane to distant points on the shores of the Pacific Ocean. The majority of these birds returned to their nests at speeds ranging from 128 to 317 miles per day.*

the Hawaiian Islands, the coast of Washington, the Marshall Islands, the Philippines, and Japan. Fourteen birds returned at speeds ranging from 128 to 317 miles per day. The longest distance covered was 4120 miles from Luzon to Midway, in 32 days.

These spectacular homing records are not typical of the average performance of wild birds. Table I includes typical homing performance found in several species of wild birds and the percentage of birds that did return home at all. Even though the displaced birds have adequate means of orientation, they may be weakened by the journey, they may require a considerable amount of time to obtain enough food, particularly in strange surroundings, and they may fall prey to one or another of the hazards that beset the lives of all wild animals. These hazards are likely to be more severe in unfamiliar surroundings.

The orientation required of a homing bird differs in certain important ways from that necessary for normal migration. In the first place, at a given season, birds of a particular species tend to migrate in approximately the same direction. To be sure, some long migration routes involve a change of flight direction at some point along the route. For example, many birds nesting in Northern Europe fly southeast or southwest to the Straits of Gibraltar or to the Middle East, and then continue into Africa in a somewhat different direction. But a bird that has been carried in a closed box several hundred miles in a direction chosen by the experimenter must not only fly the requisite distance but also must determine the direction of home. If it is a migratory bird (and most wild birds that have registered outstanding homing performance are migratory to some degree) the correct direction for a successful return from a homing experiment need bear no relation to either the spring or fall migration route. Furthermore, homing experiments are carried out during the nesting season, when the birds obviously are not migrating at all.

Wild birds, under natural conditions, are not subjected to displacement in closed boxes, and, at first thought, it would seem unnecessary for them to have a highly developed homing ability. It may well be, however, that at least somewhat similar methods of navigation are used in the two cases, and the orientation mechanisms that serve normally for migration can be pressed into service under the artificial conditions of a homing experiment. Another possibility is that homing over considerable distances is occasionally necessary when a wild bird is displaced by abnormally strong winds, either from its migration route, or from its nesting area. Indeed, many species, such as Manx shearwaters, make foraging excursions one or two hundred miles in length between spells of incubation or care of the young.

Testing Theories of Bird Navigation
by Means of Homing Experiments

The homing flights of birds offer several advantages over natural migrations for experiments that seek to elucidate the means of orientation employed. It is very difficult to observe closely the exact beginning or end of a particular bird's migration. On nights of heavy migration thousands of birds may pass over a given point; they can be heard calling or, through binoculars or telescope, they can be seen against the surface of the moon. Large numbers flying simultaneously in the same direction can be traced by radar. But there is no way of telling whether the bird whose silhouette is seen flashing across the moon began its migration ten, a hundred, or a thousand miles away, or under what conditions it departed. Nor can one tell whether a constant direction has been maintained throughout the nocturnal migratory flight.

In a homing experiment, on the other hand, the time and place of the bird's release are selected by the experimenter, and if the release point is chosen with care to avoid local distractions for the particular bird involved (such as lakes for a sea bird released inland), then much can be learned by observing its flight through binoculars during the few minutes before it dwindles to an invisible speck.

For example, homing experiments have been used to test the hypothesis that birds navigate by terrestrial magnetism. Small magnets have been attached to pigeons, and to several species of wild birds, shortly before they were released for homing flights. In some experiments the magnets were attached to the legs, in others to the head, in still others to the wings. In the last case, the rapid motion of the wings caused the permanent magnets to set up oscillating electric fields similar to those set up in a bird flying through the earth's magnetic field. In other tests of the magnetic hypothesis, the birds were placed in the fields of very powerful magnets, hundreds of times the intensity of the earth's magnetic field, immediately before release. If they possessed a magnetic sense organ sufficiently refined to be useful in navigation, one might expect some disturbance of homing performance after at least some of these treatments. But all such experiments have yielded negative results. No discernible difference was noted between the homing performance of control birds, carrying similar weights made of nonmagnetic material, and that of the experimental birds carrying magnets.

Molecular Navigation

As the data from homing experiments in the United States and Western Europe accumulated, between 1930 and 1950, a troubling thought began to occur to me and also to some other investigators. The variation in homing speed and percentage of returns, and the way in which these two measures of homing success varied with the distance of transportation, suggested that for many species, at least, a very simple and relatively disappointing explanation was consistent with the data at hand. Perhaps the birds did not possess any true ability to navigate at all! At first thought this seems a preposterous conclusion, inasmuch as some birds homed so rapidly. But a little consideration shows that it is at least theoretically possible to reconcile these apparently conflicting points of view, in two general ways.

For some species listed in Table I the percentage of returns was relatively low and decreased fairly rapidly with increasing distance of transportation. It was thinkable that these birds may have scattered in a variety of directions from the release point, and maintained relatively straight flight paths until they reached territory that was familiar to them. Many birds are known to fly over considerable distances, even during the nesting season, in order to obtain food for the young. Perhaps the birds remembered at least some major landmarks, seen on their previous migrations or foraging excursions. If so, it seemed reasonable to postulate, as a working hypothesis, that only those birds chancing to strike familiar territory were able to orient themselves correctly on the basis of remembered landmarks, and thus reach home. Since the familiar territory would be the same size for all birds from a particular colony that were released at varying distances, the angle subtended on a map by this relatively large goal area would shrink with distance of transportation. This diminishing angular size of the goal could explain why the percentage of returns fell with distance.

With the data available around 1950 it was possible to explain the recorded homing performance of most species of wild birds by assuming that the birds (1) flew at speeds that the various species were known to attain; (2) spent a reasonable number of hours per day in such flight; and (3) had no directed orientation at all. One physicist postulated that each bird flew for a reasonably long distance in one direction, then turned abruptly to fly again for some miles in any other direction chosen at random, and continued this process until it reached familiar territory. He applied to the data equations that have proved useful in describing the erratic paths along which molecules of a gas travel as they mutually jostle each other about. Such random molecular excursions can be visualized through the Brownian movement of suspended particles. This physicist came to the conclusion that the homing of many birds could be explained on this extremely simple basis. We might call this, with tongue in cheek, the "molecular" theory of bird navigation.

There remained to be explained, however, the homing of certain species, such as herring gulls, which returned in large numbers from any distance up to 500 miles. The actual release points used in experiments with these gulls are shown in Fig. 13. In a few experiments about two thirds of these gulls returned from 900 miles. While the homing speed of gulls was relatively low, these percentages of return were incompatible with the hypothesis of random scattering, and difficult to reconcile with the molecular theory. But an alternate hypothesis was easily produced to explain such cases. Perhaps these birds flew along a roughly spiral course, turning with a very large radius so as to scan wide areas in search of familiar territory. Again the quantitative results

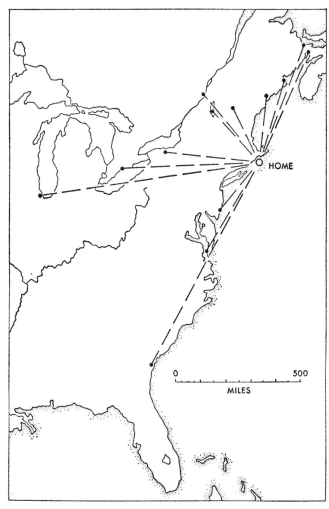

Fig. 13. Homing experiments with herring gulls. The gulls were taken from their nests on an island in Massachusetts (open circle) to the various points shown by solid dots. The great majority returned, but their speeds were highly variable.

available in 1950 could easily be reconciled with such a theory. Of course, the spiraling would require the bird to fly much greater distances than the direct course home, but the homing times of herring gulls were almost always quite long enough to allow for flight over many times the direct homeward route.

While these hypotheses were disquieting in a number of ways, they had the advantage of explaining the observed facts of homing without assuming either that the birds possessed specialized sensory capacities, or that they were responding to some environmental signals beyond human recognition. If the behavior of homing birds could be described by the same equations that accounted for the random motions of molecules, this explanation had a very great appeal, especially for physical scientists who tend temperamentally to feel uncomfortable when faced with the complexities, and the apparent mysteries, of biology.

Chapter 6

BIRD WATCHING FROM AIRPLANES

Twenty years ago the results of homing experiments had reached a tantalizing state. All the available data could be accounted for if one assumed only the simplest sort of exploratory behavior, or even if one applied the "molecular" hypothesis that treated homing birds like jostling gas molecules. Yet it hardly seemed credible that birds would behave so inefficiently. What was needed, obviously, was to trace their actual flight paths. In 1940, I decided to approach this problem in the most forthright manner possible—by learning to fly a light airplane and following some homing birds to see where they really went. This experience convinced me that the behavior of birds on the wing can be appreciated much better if one accompanies them into their own medium. Hence this chapter will include accounts of several episodes that may help to convey a little of this feeling.

Gulls and Gannets

The first species selected was the herring gull, but before the war made such work impossible, there was time only to develop the technique to a slight degree and to follow a few gulls from ten to fifteen miles at the most. These birds did not start at all directly toward home, the seemingly random departures tended to support the molecular hypothesis. But herring gulls are adept at riding updrafts. Once I watched five gulls in succession climb rapidly under cumulus clouds and drift downwind for several miles at about thirty miles per hour. Since their home was 104 miles downwind, soaring at their speed should have carried them home in three to four hours. Indeed, one gull did reach its nest three and a half hours after release. The gulls' heavy reliance on updrafts contributed another large variable to an already complex phenomenon. Furthermore, they were known to range widely not only along the coast, but also inland. Unless they were taken hundreds of miles from home, one could not really be certain that in a lifetime which may last twenty years or more they might not have visited the vicinity of the release point and remembered some landmark.

After the war I turned to gannets for my airplane observations of homing experiments. Experience with gulls had shown that white birds were easier to see against almost all kinds of terrain than those of any other color. Gannets were not only white but twice as large as herring gulls, with a wingspread of about five feet. Furthermore, they were strictly marine birds, never seen inland except after heavy storms. Another advantage was that gannets are heavy enough to have great difficulty in taking off unless they have a strong head wind, or a cliff from which to launch themselves into the air. On a completely calm day they cannot take off from the water. It therefore seemed likely that, if released inland, they would continue seeking home once they were in the air and would avoid landing on any but the largest lakes.

It was not until the summer of 1947 that R. J.

Hock and I could make suitable arrangements for a number of homing experiments with gannets nesting on Bonaventure Island, off the Gaspé Peninsula, in the Gulf of St. Lawrence. Most of the birds were released at the airport in Caribou, Maine, in the valley of the St. John's River, which flows

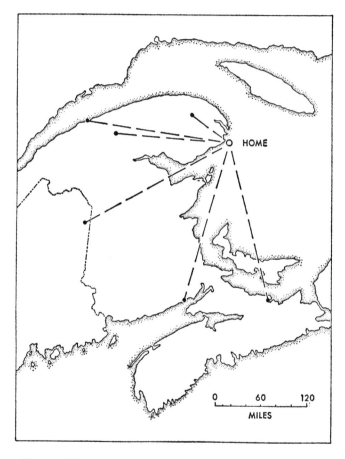

Fig. 14. Homing experiments with gannets. The birds' nests were on Bonaventure Island in the Gulf of St. Lawrence (open circle); they returned from the points indicated by solid dots.

generally southward to the Bay of Fundy. (See Fig. 14.) The reason for choosing this release point was that it lay approximately 100 miles from salt water in three directions—northwest, northeast, and southeast. I tried to anticipate what a gannet might do when it suddenly found itself over the potato fields of northern Maine, and it seemed possible that the drainage pattern might guide the bird downstream along the first river it encountered. Such a tactic would eventually bring any sea bird to the ocean. Working down the valley of the St. John's River, or simply avoiding the wooded hills that were visible on clear days from Caribou in three directions, would tend to take the gannet south rather than east northeast, the correct bearing of the home island.

Preliminary experiments showed that while gannets were not quite as faithful homers as herring gulls, they returned at an average speed of ninety miles per day from Caribou to Bonaventure Island. Ten out of sixteen birds released at Caribou on several different days in good summer weather did return, the fastest in 24 and the slowest in 101 hours. This corresponds to homing speeds of 213 and 51 miles per day. Gannets, like gulls, seemed to enjoy soaring, and commonly devoted long periods to it. But aside from their seeking out cumulus clouds, their behavior did not seem as directly influenced by updrafts as had that of herring gulls. Once underway, the gannets usually flew continuously, with considerable determination, for many miles. They might detour slightly at times to catch a good updraft, if one could be located by the cumulus cloud forming above it, but for the most part they flapped doggedly ahead.

We used high-wing monoplanes slightly larger

than the smallest Piper Cubs popular at small airports for training and pleasure flying. I usually flew the plane, and an observer sat behind me. The design of the plane allowed an unobstructed view on each side of the engine over an angular range of slightly more than ninety degrees vertically and horizontally. It was essential to maneuver so that the bird always remained in either the right or left cone of visibility. To minimize any disturbance of the bird's behavior we stayed as far away as we could without losing sight of our bird. Since the airplane flew faster than either gulls or gannets, some sort of maneuvering was necessary to keep the bird in view, and a simple, slow circling proved the most effective. When following a gannet we tried to hold the plane about 2000 feet to one side of the bird and 1500 feet above. Even on gentle turns it is necessary to bank a plane, and unless we were considerably above the bird, the dipped inside wing would cut it off from view. Safety in such low-speed maneuvering also called for considerable altitude.

While some birds, geese among them, are obviously frightened when a plane comes within a mile or two, herring gulls, pigeons, and gannets showed no sign of disturbance unless we approached within one or two hundred feet. The homing performance (speed and percentage of returns) of gannets we escorted for several hours did not differ from the usual performance data for the species. Hence, we were reasonably confident that the airplane itself was not introducing a serious distortion in the homing behavior.

Nine gannets were followed for appreciable periods of homing flight, and their flight paths are shown in Fig. 15. The shortest observation was for 25 miles, the longest for 230 miles. The time the birds were

held under observation varied from just over an hour
to more than nine hours. The pilot had to maneuver
almost continuously in order to keep a proper distance and at the same time hold the bird in a

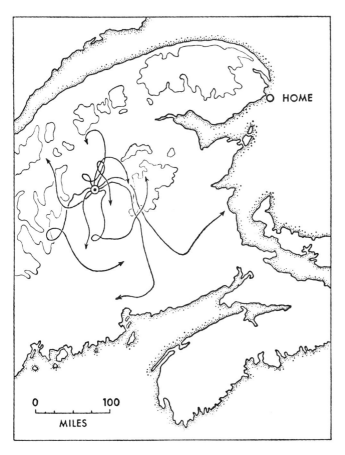

Fig. 15. Flight paths of homing gannets traced by airplane.
The release point at Caribou, Maine, was chosen so that hills
to the west, north, and east might encourage flights to the
south, but the actual homing flights showed little if any influence of topography. Contour lines indicate elevation of
1000 ft.

favorable position of observation. The line of sight to the bird had to be clear of the undersurface of the wing and the landing gear. In a quick turn the bird might slip so far ahead of the airplane that the propeller and engine cowling would hide it from view, or it might drop too far to the rear and vanish with disconcerting suddenness.

The gannets flew about thirty-five miles per hour, the airplane roughly sixty miles per hour. This difference in speed required one circle every few minutes, and the slow circling allowed time for us to observe landmarks and plot our approximate position on a map. Much of the territory around Caribou is heavily wooded, and while there are many streams and lakes, there was a considerable risk that the bird might fly over areas devoid of landmarks included on the map, and that we might not be able to plot the course. Against this contingency I carried a motion-picture camera on my lap to photograph the wooded terrain. The hope was that we could later return to the area with prints from the motion pictures and thus re-establish where we had been. Fortunately, I never had to use this desperate measure of ex post facto navigation, and, indeed, I had no great confidence in the idea.

Many improvements could have been introduced in these experiments if we had had more to work with. It would have been much better for both people in the plane to have known how to fly, so that one could relieve the other. Of course, if one were to do such experiments in the 1960s, radio aids to navigation, not so available in 1947, would be extremely helpful.

Spectacular Soaring Demonstrations

Our flights produced some interesting experiences apart from the information we gathered on homeward navigation. Two gannets afforded exciting observations of spectacular soaring ability. They found strong updrafts under sizable cumulus clouds and proceeded to ride these ascending currents of air with an upward velocity the airplane could match only at full throttle. The reason for our ignominious performance was that we were circling outside the updraft, probably most of the time in the corresponding downdraft, while the gannet was expertly maneuvering to stay in the most rapidly rising currents of air. Furthermore, an airplane has considerably less lift while turning, because it must necessarily be banked and hence presents a smaller wing surface area to oppose gravity. The end result of both these vertical pursuits was that the airplane circling several hundred feet above the bird reached the base of the cumulus cloud, and was unable to climb higher without entering the overcast and losing sight of the bird (as well as violating every regulation in the book, inasmuch as neither the airplane nor I was prepared for instrument flying).

My problems did not deter the gannets, however, and they continued happily climbing until, in one case, I lost sight of a bird behind the inner wing, and in the other instance I managed to see it disappear behind a wisp of cloud that formed part of the irregular cloud base. It would be most interesting to know what a gannet or other bird does when an updraft carries it inside the overcast. Did my gannets descend again fairly soon, but escape my notice because of the limited visibility from the airplane? Or did they continue to climb to the top of the cumulus cloud and thus find themselves at the highest possible altitude, but probably cut off from a view of the ground by the thick layer of cloud?

On the second occasion when this disappearing act was obviously imminent, I decided to try an unorthodox maneuver in hopes of keeping the gannet in view a little longer. The customary banking of an airplane during a turn prevents the sideways skid which would occur on turns without banking. First, I tried to make skidding turns by turning the rudder without tilting the plane, but, as expected, this simply caused my plane to slide laterally away from the gannet, and even if such skidding turns had been successful, I would have lost the bird from view as soon as it had climbed to my own altitude. Exasperated, I decided to try banking the plane in the opposite direction. The wing area would resist skidding just as well, and perhaps I could keep a clear inward view toward the center of my turning circle where the gannet was soaring up to the cloud base. If something went wrong I had 5000 feet of altitude to get the plane back under control. In one respect this innovation was successful; it was possible to circle to the left with the plane banked to the right, and even to keep on climbing with the gannet. But an alarming vibration of the airplane set in at once. Since I could not think what could be its cause, or probable outcome, I returned discreetly to conventional flying while my gannet escaped from view, leaving me as ignorant, and fascinated, as ever about its behavior at the cloud base. Later I was informed that the vibration was caused by unequal speed of the air flowing past the propeller on the two sides of its circle of rotation, an unhealthy condition at best for a light airplane not designed for acrobatic flight.

Four of the gannets followed over considerable distances returned to the home island after periods ranging from forty-five to seventy-five hours, and their flight paths illustrate the actual courses flown by homing gannets over totally unfamiliar inland terrain. As can be seen in Fig. 15, the flight paths suggest no tendency to choose the correct direction. Indeed, the behavior of these gannets was consistent with the hypothesis of spiral exploration discussed in the previous chapter. There is reason to suppose that much if not all the coastline shown in Fig. 15 would be familiar to these gannets, whereas no territory inland from the coast would have been visited by these birds in any ordinary circumstances.

In only one case were we able to follow a gannet all the way from Caribou to the coast. But as luck would have it, the day on which this most successful flight occurred was marred at the outset by an electric power failure in the town and airport of Caribou. The regular fuel tanks of the airplane had been filled the previous night, and they held enough for approximately seven hours' flying time at the reduced speed of our maneuverings. The special auxiliary tank, which extended the flying time to eleven hours, could not be filled that morning because all fuel pumps were electrically operated. All other conditions were ideal, and any delay would have necessitated holding the bird in captivity two days instead of one, with almost certain deterioration in its physical condition. We therefore decided to release the gannet and follow it as best we could with fuel for only seven hours. Everything else went better that day than in any other flight. The gannet found favorable updrafts part of the time but never climbed so close to the cloud base that we lost it. It proceeded, generally eastward, along the course shown in Fig. 15, ignoring rivers and lakes, including even the very large Grand Lake in central New Brunswick. Knowing that the bird was approaching the coast, we were most interested to see what its reaction would be when it reached salt water. But our fuel was running dangerously low, even before we could see the Gulf of St. Lawrence, and we were some distance from any airport. Almost at the same moment the ocean became dimly visible and the gannet began a steep descent from its previous cruising altitude of 2000 or 3000 feet. But we already had little enough fuel, even supposing that all went well in a direct flight to the nearest airport. Considerations of common sense compelled us regretfully to break off the observation just as the gannet was approaching its native element.

There was a further practical complication in that we had flown from Caribou over the international border more than one hundred miles into Canada. The rules governing such flights were unpleasantly clear and unequivocal. While one was allowed to fly a few miles across the boundary without special permission, provided one did not land, any flight of more than fifty miles into Canada, or any landing there, required prior approval of a customs officer. Arrangements had to be made so that a customs official was on hand when the airplane arrived, in order to clear its entry. Furthermore, because of the delayed start while we were vainly waiting for the Caribou Power and Light Company to restore its services, there was barely enough daylight for the return flight to Caribou. Since my experience with flying at night was entirely inadequate, I made a hasty and furtive landing at the nearest airport, re-

filled our tanks, and fled unheralded back across the
international boundary.

Apparent Support for the "Molecular" Hypothesis

What we observed while following the actual
flights of herring gulls and gannets tended to support the "molecular" hypothesis that homing from
unfamiliar territory was not accurately directed, but
succeeded eventually because of exploratory flights.
Scientists always prefer the simplest theory consistent with the known facts, and this hard-boiled
explanation had the advantage of demanding nothing more of the birds than an ability to recognize
landmarks, coupled with the perseverance needed to
search many hours or days for familiar territory. Here
was a situation that seemed clearly to call for application of "Occam's razor," so named for the
fourteenth-century Franciscan theologian and
philosopher William of Occam (or Ockham). He
argued eloquently that one always should credit the
simplest explanation available for any phenomenon,
the one requiring the fewest and least complex
philosophical assumptions. But living organisms
often surprise us in their subtle complexities, and
shortly after 1950 new experiments threw quite a
different light on the whole question. Despite its
tidy respectability, after pruning by Occam's razor,
the hard-boiled "molecular" explanation of bird
navigation has been forced to give way before new
evidence, which is the subject matter of the following chapter.

Chapter 7

EXPERIMENTAL ANALYSIS
OF BIRD NAVIGATION

While the evidence summarized in the previous chapter did suggest that herring gulls and gannets managed to home without any highly developed ability to select the home direction, there was no compelling reason to apply this explanation to all birds. Suspecting that other kinds of birds might well display superior powers of navigation, Harold B. Hitchcock and I turned to homing pigeons in an attempt to learn something about their navigation by tracing their flight paths from small airplanes. Hitchcock's pigeons performed better than mine, and therefore Fig. 16 shows some of the longer flight paths which he succeeded in following. It is apparent that these pigeons deviated considerably from the direct path between release point and home, but they did better than one would expect from the molecular theory of bird orientation.

In the early 1950s two important advances were achieved in the explanation of bird navigation, one by Geoffrey V. T. Matthews at Cambridge University, in England, the other by Gustav Kramer at Wilhelmshaven, Germany. Matthews worked mainly by observing and analyzing the initial headings of birds in homing experiments. Kramer's experiments (to be discussed later) included not only observation of the initial headings of released homing pigeons but also study of the migration restlessness (described in Chapter 3) of a few selected, hand-raised, and caged wild birds.

For many years pigeon racers had maintained that on release homing pigeons often would fly straight off toward home. But in most cases racing pigeons are released in large flocks containing many birds experienced with the particular release point and the homeward route. A reasonable explanation, in

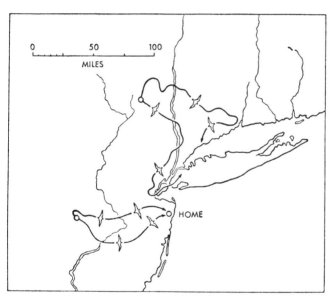

Fig. 16. *Flight paths of two flocks of racing pigeons followed by Hitchcock. In the training program birds were released to the west of the loft near the coast of New Jersey, and then followed during two flights from an unfamiliar release point farther west, in Pennsylvania. They were next taken north into totally strange territory. Again, they started east or southeast, but eventually turned south; most reached their loft even though airplane observations could not be continued all the way.*

the absence of other information, was that these experienced birds remembered the correct route and were followed by the others. In many homing experiments the immediate surroundings at the release point sometimes seemed to have an effect on the bird's choice of direction for its first flight. For instance, sea birds released inland tended to head for any body of water, while pigeons were likely to display some interest in towns or farms.

Matthews' Observations of Initial Headings

Matthews made a special effort to select release points at the centers of large open spaces of level terrain with a clear view in all directions. Every effort was made to avoid local features of topography that might influence the direction in which the birds would fly when first set free. To guard against bias it was necessary, of course, to vary the direction in which the birds' heads were pointed at the actual moment of release. This pointing was done in an irregular fashion, perhaps with the first bird headed north, the next southwest, the third east, and so forth.

In his first experiments Matthews trained pigeons by releasing them at increasingly distant points that all lay approximately on a single line. This training direction extended north northwest from Cambridge. As in all properly conducted homing experiments, the birds were transported in closed boxes, usually by automobile or light truck, and they had no opportunity to see anything of the countryside. When thus released in unknown territory in an unknown direction from home, many of the pigeons headed initially in their customary southward homing direction. These birds, it was apparent, had established a

preferred direction of flight which they chose regardless of the release point. When taken to a new release point on the training line, several miles farther
north than their previous starting point, the pigeons
again displayed a clear ability to fly south—toward
Cambridge.

Naturally, Matthews wanted to find out on what
basis this choice was made. By releasing pigeons under a variety of weather conditions, he made the important discovery that the initial headings were far
less consistent when the sky was heavily enough overcast to hide the sun. While there was the usual variability from day to day, and from one pigeon to another, the choice of an appropriate direction seemed
to be greatly impeded, if not rendered altogether impossible, when the pigeon could not see the sun. It
had long ago been suggested that in pigeon races,
which always start early in the morning, the birds
might take advantage of the sun's position to determine the appropriate homeward direction in which
to fly. But it seemed incredible that a bird might use
the sun as a "compass" to determine directions at
any time of day. Matthews was the first to consider
this possibility seriously enough to carry out experiments that tested it directly.

In one series of releases along in the training direction north northwest from the pigeon's home
aloft near Cambridge the releases were made always
in the middle of the day when the sun was nearly at
its highest position in the sky. It was thus approximately south of the birds when they made their initial choices of flight direction. The simplest explanation for these directional choices would be that the
bird had learned to fly toward the sun. But these pigeons did not merely start in the correct direction;
many of them also reached home at relatively high

speeds. A simple tendency to head toward the sun would have caused them to turn more and more to the west as the afternoon wore on, and by evening they would have been flying almost directly away from home.

Matthews investigated the matter further by experimenting with pigeons which had been released several times at increasing distances up to seventy-eight miles along the training line, always at approximately midday. Dividing the flock into three groups, he transported them forty-nine miles farther (that is, 127 miles from home) in the same direction. One group was now released early in the morning, another late in the afternoon, and a control group in the middle of the day as usual. On the average the experimentals were released six hours earlier or later than the controls. The result was that all three groups chose the homeward direction with equal accuracy. It made no difference that the sun was roughly ninety degrees east or west of the position where they might have come to expect it. The results of this important experiment are illustrated in Fig. 17, which shows not only the similarity of the headings when birds were released early in the morning, at midday, and late in the afternoon, but also the general accuracy of these homeward headings.

After Matthews had completed the experiments just described, he did not rest content with the demonstration that his pigeons could use sun-position for headings. He continued and extended the same basic type of experiment and soon found that some groups of pigeons performed in a still more interesting manner. After considerable experience at distances up to fifty miles or more, along a single training line, they were released in unknown territory in some quite different direction. In the most important experiments the training direction was again north northwest, but the crucial tests were made at release points to the south or west. Again there was considerable variability, and a few of the birds flew roughly south as they had done on previous occasions when released along the training line. Others scattered in many other directions. But a considerable majority flew off more or less toward home. This was true whether home was north or east. After a number of experiments had been completed, it became clear that many of the pigeons could select ap-

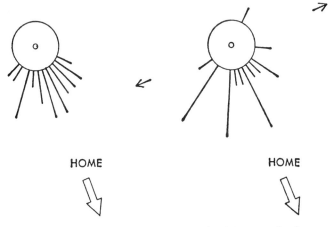

HOME   HOME

Fig. 17. Initial headings of pigeons trained by several releases at midday. Each release was farther north northwest from the home loft, the release previous to this one having been at 78 miles. At left, the initial headings of birds released at 127 miles (in totally strange territory) around noon. At right, the headings of others from the same group released at the same place in early morning and late afternoon. The length of each line is proportional to the number of pigeons starting in the direction indicated; the shortest lines signify one bird. Small arrows show the directions these birds would be expected to take if they had flown at the same angle to the sun as in the training flights.

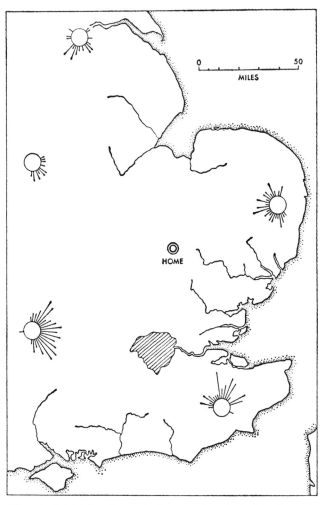

Fig. 18. *Initial headings of pigeons released by Matthews in unfamiliar territory after transportation in five different directions from the home loft (double circle). Each line radiating from circle around a release point is proportional in length to the number of pigeons flying in the direction indicated. The shortest lines represent one bird. Training directions in previous flights were different, in most cases, from the direction of transportation for this experiment.*

proximately the homeward direction whether or not it coincided with that of the previous training flights. Examples of several experiments of this type are shown in Fig. 18. Despite the considerable variability, the choices made by the great majority of birds lay far too close to the home direction to be explained by random scattering, or by any of the explanations that appeared applicable to the gannets and other homing experiments prior to those of Matthews.

In spite of the definite demonstration that pigeons could establish homeward orientation within the first few minutes after release in unknown territory, the interpretation of such initial headings is complicated by the fact that in every experiment at least two types of orientation must be considered as distinct possibilities. Are the birds only choosing a roughly constant direction (in the case of pigeons usually a training direction)? Or will they choose the true home direction regardless of which way they may have been displaced into the unknown surroundings of the release point?

If a bird is taken in the direction opposite to a preferred flight direction, accurate homeward headings do not establish which type of orientation is involved. Other experimenters with pigeons have found that their birds exhibit a constant directional tendency in their initial headings independent of the direction they have been carried from home. For example, pigeons trained to home to Wilhelmshaven near the northern coast of Germany, where most of Kramer's experiments were conducted, show a northward tendency in their initial headings. Only after releases in two or more different directions can one determine with reasonable confidence whether a unidirectional tendency is at work or true homeward orientation.

The Initial Headings of Manx Shearwaters

In addition to his work with domestic homing pigeons, Matthews also experimented with Manx shearwaters, birds which had already been shown to have a most impressive homing ability. It was Manx shearwaters which returned from Venice and Boston to their nesting island off the coast of Wales at average speeds of 265 and 244 miles per day. Matthews transported shearwaters from the same island to various points inland in England where the terrain was relatively uniform and an observer on a good vantage point (such as the library tower at Cambridge University) could follow the birds with binoculars for considerable distances. Again there was variability, some shearwaters choosing directions far removed from the correct homeward path. But, on the average, shearwaters performed as well or better than the best homing pigeons, and a tabulation of results from several inland release points in England showed beyond doubt that most of the shearwaters were choosing approximately the correct homeward direction and were not scattering in anything approaching a random pattern.

These experiments obviously carried the whole problem of homing in birds into a new phase. Other investigators tried essentially the same methods on other species, but no other bird has yet been found which shows as consistently accurate initial headings as the homing pigeons, used by Matthews and Kramer, or the Manx shearwater. Other not-so-select strains of homing pigeons give results that are far less convincing. Nevertheless, the fact that some birds can consistently choose approximately the correct homeward direction, and demonstrate the correctness of this choice in their initial headings, means that we are dealing with an ability to orient which makes itself manifest within a very short time after release. No one can yet say with certainty whether it is only a few species of wild birds, and only the best strains of homing pigeon, that possess this ability, or whether it is widespread among birds but has only been demonstrated in a few favorable cases.

Kramer's Experiments with Orientation Cages

At about the time when Matthews was conducting the experiments described, Gustav Kramer and his associates were experimenting with orientation exhibited under more strictly controlled conditions in experimental orientation cages. In his first investigations Kramer used starlings which had been raised by hand from an early age, and were very tame. They were placed in circular cages provided with perches both at the center and also around the edges. As had been discovered many years earlier, such migratory birds exhibit a pronounced restlessness at the season when they would normally undertake long migratory flights. As explained at the end of Chapter 3, this migratory restlessness occurs primarily at night, when many small birds migrate, but Kramer's clearest experiments were carried out in daylight, employing a few well-tamed starlings, which are diurnal migrants.

The starlings were placed in a circular cage which was carefully built so as to be completely symmetrical; its inner surface had an identical appearance when the bird faced in any direction. The observer lay on his back on the floor and watched the birds from below through the transparent plastic floor of the orientation cage. One tame and co-operative starling tended in the spring to fly back and forth from the central perch to the edge of the cage primarily in a northwesterly direction. This direction corresponded roughly to the normal northeasterly spring migration of starlings in the area where it had been taken from its nest in its first few days of life.

On what basis did this starling make its directional choice? In addition to being symmetrical, the orientation cage was arranged so that it could be rotated, and when it was turned from time to time during the period of migratory restlessness the starling continued to head in the same general direction, regardless of which wall of its cage had to be approached in order to head northwest from the central perch. The next step was to surround the orientation cage with a uniform, opaque fence or shield, which also could be rotated about the center of the cage. This screen cut off any view of the surrounding landmarks, but allowed the starling to see most of the sky. Again it made no difference how the screen was rotated. Clearly the bird was not responding to local landmarks. Kramer also established that the bird and orientation cage could be moved about from point to point in the vicinity of Wilhelmshaven, without occasioning any essential change in the directional tendency.

What aspect of the sky provided the directional information underlying these choices of the correct direction for fall migration? One possible directional cue was the sun. For the decisive experiment Kramer enclosed the orientation cage in a further opaque structure which excluded all view of the outside world except what the bird could see through six large windows. This starling continued to show the

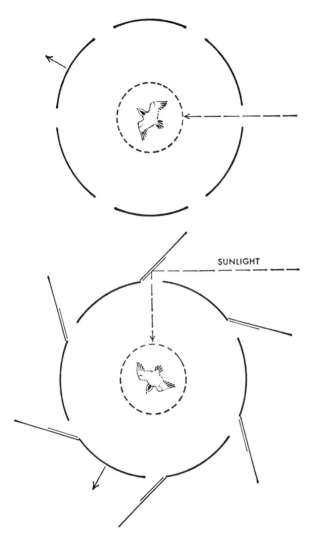

SUNLIGHT

Fig. 19. *Directional headings of Kramer's starling during migration restlessness. In upper diagram the bird heads approximately northwest when the sky is directly visible through six windows in its otherwise opaque enclosure. In the lower diagram each window has an opaque screen and a mirror which reflects the light so that it appears to enter at approximately 90 degrees from its normal direction; the bird now heads southwest.*

same northwesterly heading when it could see the
sun and blue sky through one or more of these windows. The next step was to equip the windows with
shutters whose inner surfaces had large plane mirrors, as shown in Fig. 19. These shutters were so arranged that the light reflected into the window by
the mirror came from ninety degrees to the right or
left of the window itself. In other words, looking out
the window, the bird saw reflected in the mirror an
area of sky ninety degrees to one side of the sky directly beyond the window. The starling continued to
show strong directional preferences, but now the
bird maintained the same angle relative to the light
reflected from the mirrors as it had to the normal
sunlight. Thus, when the mirrors showed the sky
ninety degrees to the right of what normally would
have been visible through each window, the starling
headed approximately ninety degrees to the right of
its previous heading. The opposite result occurred
when the windows were so arranged that the sunlight reached the bird from ninety degrees to the
left of its normal angle of incidence.

When the sky was completely overcast the headings were virtually random, but when the sun became visible a few hours later the bird again chose
approximately the northwest direction for its headings. Kramer thus showed that one particularly co-
operative starling was able to choose a consistent
direction in his circular orientation cage, and, furthermore, that this direction was determined with reference to the sun. The correspondence between this
experiment and Matthews' studies of initial flight
direction in pigeons was obvious and compelling.

Food-seeking Experiments with Starlings

It was extremely difficult to obtain other starlings as co-operative as the one used in the experiments just described. Kramer therefore turned to a different experimental procedure which did not involve reliance on the bird's spontaneous choice of migratory direction. Instead, he trained starlings, and later other birds as well, to obtain food by seeking it in a particular direction from inside a circular orientation cage. The cage differed from the one used for the previously described experiments in that several identical food chambers were equally spaced around the outside edge of the cage. Each food container was covered on the side facing the bird with a slotted rubber membrane. It was impossible for the bird to see through the slot, and only the experimenter knew which of the feeders actually contained food. Nevertheless, starlings and other birds learned to thrust their bills through the slot and pick up any grain that was available. As in the other experiments the array of food chambers, the circular cage, the perches inside it, and the surrounding circular screen that hid the local landmarks could all be rotated independently. At irregular intervals throughout the training and the later critical tests these parts of the apparatus were rotated in different ways. Thus the birds had no way to tell from the immediate features of the orientation cage which feeder contained food. When the clear sky and sun were visible to a starling in such an orientation cage it would learn in the course of two or three weeks, if it was a co-operative and reliable bird, to reach for food in the correct direction with errors seldom exceeding thirty degrees.

If anyone had discussed this experiment with biologists or other scientists in advance of Kramer's actually carrying it out, it would have been agreed rather generally that a bird could be expected to learn to find the correct feeder, provided the feeder remained in the direction of the sun or at some constant angle to it. It would have been very difficult to believe, however, that a bird could easily and rapidly compensate for the sun's movement through the sky. At first thought this accomplishment certainly seems much more difficult than learning to seek food at a fixed angle relative to a very bright and conspicuous object such as the sun.

For critical tests Kramer always used birds that had already learned to make accurate directional choices under some standard test condition. The bird was then presented with a set of empty feeders, and its first several choices were carefully noted. It was necessary that the feeders be empty in order that the bird not be motivated by its success, or lack thereof, in the initial trials of the critical experiment. Fortunately the birds learned their initial lesson so well that they would persist for dozens of trials even when unrewarded. In one of the clearest of several experiments Kramer trained a starling to seek food from the feeder to the east of its cage. All its training over a twenty-day period occurred between seven and eight in the morning. For the critical test the orientation cage and bird were moved seven miles to an unfamiliar environment and the tests were conducted at 5:45 in the afternoon. In this experiment six feeders were available, equally spaced sixty degrees apart. In twelve tests with empty feeders seven choices were made at the east feeder and four others at the two adjacent feeders, sixty degrees right and left of the correct one. Only one out of twelve choices was made at the west feeder. During the original training in the morning the sun had been slightly to the right of the east feeder; during the critical tests in the late afternoon it was almost directly behind the west feeder. If the starling had simply learned to head in a fixed direction relative to the sun, its choices in the afternoon should have been to the west.

Probably Kramer himself was astonished at this result, and one of his assistants attempted the opposite experiment of training another starling always to seek food in the feeder closest to the sun. In this case, the training had to be scattered through all hours of the day. Seven weeks of strenuous training effort was necessary before this starling (appropriately named "Heliotrope") became moderately reliable at choosing the feeder nearest to the sun. Another starling trained at the same time as Heliotrope to seek food in a fixed direction had learned this lesson after only ten days.

Sun-Compass Orientation

Kramer had thus shown that at least some of his hand-reared starlings learned to seek for food in a fixed-compass direction rather more quickly than they learned to solve what seems to us a far easier problem—to look for food in the feeder nearest to the sun. Taken together with the results of the previously described experiments of Matthews, this demonstration indicates very strongly that birds do indeed know how to compensate for the apparent movement of the sun across the sky. This type of orientation has come to be called "sun-compass orientation," meaning simply that the bird or other animal uses the sun's position to determine a particular direction. Essential to sun-compass orientation is the

ability to choose an angle of travel relative to the
sun's position—an angle that must change continuously through the hours of the day as the sun's position moves across the sky. At first thought this
seemed an astonishing achievement for an animal.
Indeed, for a long time biologists refused even to
consider it seriously. But in recent years careful observations and experiments have shown that sun-
compass orientation is exhibited not only by birds
but by many small invertebrates, such as shrimps,
spiders, insects, and also by fishes, and lizards.

Biological Clocks of Animals, Plants, and Microbes

The ability of birds to correct for the sun's changing position throughout the day can be understood
better when it is considered together with other biological phenomena that demonstrate the operation
of time-keeping mechanisms within the bodies of animals and plants. These timing mechanisms often
make their presence known through what are called
endogenous activity rhythms. Almost every animal
and plant in its normal life changes its activities from
day to night. Flowers open during the daytime; animals sleep by day and are active at night, or vice
versa. These changing patterns of activity are most
easily explained as direct responses to the changing
physical environment. To consider birds, most of
which are diurnal, one can most reasonably suppose
that they go to sleep as darkness falls and awaken
again with the coming of daylight.

Strict application of Occam's razor makes this explanation seem so logical and obvious that some
have thought it a waste of time to consider any other
possibility. But biologists have wondered from time
to time what would happen to animals and plants

that were maintained under constant environmental conditions. Suppose we keep an animal in a laboratory room where temperature, humidity, sound, and light are held as constant as practicable to remove all diurnal variations. Since there is nothing to distinguish day from night, we might well predict that the animal's activity would either be constant or would vary in some irregular fashion bearing no relation whatever to the normal twenty-four-hour cycle of day and night. With some kinds of animals this is indeed so, but for a large majority careful recording of some aspect of the animal's activity shows that a twenty-four-hour cycle persists even under virtually constant conditions. Birds held in total darkness show an almost complete lack of activity, but a constant, relatively dim illumination allows the bird to see its way about in its cage, find food and water, and generally to remain in a reasonably contented condition. The easiest form of activity to measure is the total muscular movement of a bird. This measuring is commonly done by suspending the cage from light springs or rubber bands so that every step or hop the bird takes causes some jiggling, which can easily be registered. If such a recording is now examined over a period of many days, one very often finds a regular twenty-four-hour cycle of the bird's movement, even under the most constant conditions that can be devised. Commonly there is a major period of activity in the early morning and another late in the evening.

Small nocturnal mammals, such as rats and mice, have been studied far more extensively than birds, and we have no reason to believe that their biological clocks differ in any basic fashion. There is so much more accurate data available from studies of nocturnal rodents that we can obtain a clearer view of biological clocks in general by turning briefly from birds to consider experiments on a common type of small wild mouse, the deer mouse, or white-footed mouse. These attractive little mammals are strictly nocturnal under natural conditions; they remain at rest in burrows or other shelters during the day and begin to move about only after darkness. They live contentedly in small cages in a laboratory, particularly if these cages are provided with exercise wheels in which the deer mice can run for as long as they wish. An exercise wheel is a convenient mechanism for recording this running activity.

If a deer mouse that is well established in such a cage under the normal cycle of day or night is now subjected to constant conditions, including continuous darkness, it continues to run about as before. These animals have no difficulty moving about their cages in the dark or in finding food, which in such experiments is provided in abundance. Almost every deer mouse shows at least a rough sort of twenty-four-hour cycle in its running activity. Some individuals are remarkably precise in the timing of this activity, and the results of an experiment by K. S. Rawson on one such accurately timed animal are shown in Fig. 20. After several days of normal day and night, this deer mouse was kept in constant darkness for the remainder of the experimental period shown in the figure. As can be seen at a glance, its activity continued to be cyclic, and energetic running in the exercise wheel began quite suddenly at approximately the normal time every evening. The onset of this running activity occurred with a remarkable precision at a predictable time plus or minus two or three minutes.

Many plants and microorganisms have been found to exhibit similar diurnal activity rhythms under constant conditions. The diurnal cycle of opening and closing of flowers also persists in many cases under constant conditions. A one-celled organism called Euglena that swims vigorously by means of a long flagellum or whiplike appendage, moves into a beam of light which is turned on periodically to illuminate part of the jar containing a culture of

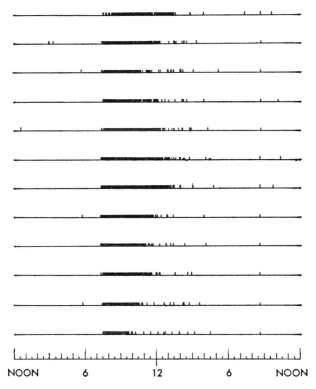

Fig. 20. Activity records of a deer mouse living under controlled laboratory conditions. Each line represents one day, and a vertical deflection of the recording pen corresponds to several revolutions of the exercise wheel; the continuous solid bar signifies vigorous running. Note that the heavy activity begins at almost exactly the same time each evening.

thousands of Euglena cells. But the speed of this attraction to the lighted area, and the proportion of Euglena that come to the light at all, vary greatly on a twenty-four-hour cycle which persists under constant conditions. Another example of a twenty-four-hour "clock" in a microorganism is the emission of light by certain one-celled marine algae when they are disturbed. The amount of light emitted has been shown to vary periodically on approximately a twenty-four-hour schedule, even under constant laboratory conditions. Still more remarkable, from the point of view of the biochemist, is the fact that many of these biological clocks keep almost the same time at high and low temperatures—within the range that the organism can survive at all. Almost all chemical reactions are speeded by a rise in temperature; typically their rate doubles every time the temperature rises by 10° C. But biological clocks, which are presumably based on biochemical processes, are temperature-compensated.

Timing by Physiological Processes

Some biologists have doubted that such precision and independence of temperature are possible for a mechanism confined exclusively within the bodies of animals, plants, and microbes. They have asked how one can be really certain that all influences from the outside world have been excluded from the constant-temperature room in a laboratory where biological clocks are studied. Three sorts of evidence have satisfied most biologists that such a concern is no longer necessary. Animals that have displayed an endogenous activity rhythm under constant conditions have been transported by airplane from Paris to New York or New York to California to another laboratory where the experiment was repeated immediately. The result has been that the biological clock continued to run on the same time as before in the previous environment. If the clock were set by local conditions, leaking into the laboratory from the outside world, one would expect a nocturnal animal to start its activity at the time of local evening, rather than the evening at its place of origin hundreds of miles to the east. A second type of evidence is that the biological clocks seldom keep perfect time, and the time from the start of one activity period to the next is never exactly twenty-four hours, but usually slightly less. This explains why in Fig. 20 the onset of the deer mouse's activity occurred progressively earlier with respect to the outside day and night. This particular record was chosen for the constancy of the mouse's timing, but most animals held under constant conditions drift out of phase with the outside world more rapidly. In time, such an isolated animal may be active at a time twelve hours different from its normal period of wakefulness. If external factors controlled the timing, the activity should not deviate substantially from the outside day and night.

The third type of evidence that biological clocks really are timed by physiological processes inside the animal comes from experiments that have proved very helpful also in explaining how birds use such internal time-keeping processes in sun-compass orientation. The basic procedure is to reset the clock by keeping the animal for several days in a closed chamber where electric lights provide a twenty-four-hour cycle of light and darkness that is out of phase with the real day and night of the outside world. For instance, in a given experiment, the lights might be turned on at midnight, and off again at noon, at a season or latitude having a natural day beginning with sunrise at six A.M. and ending with sunset at six P.M. The birds kept on such a schedule come to adjust their activities within a few days to fit the new routine, and they would begin and end their daily activities six hours earlier than a control group exposed to the natural daylight. For birds and most other vertebrate animals light is the most effective environmental factor that can be used to reset the biological clocks in this fashion.

After a bird has been thus reset, it can be subjected to constant dim illumination. The result is that it continues to show an activity cycle timed like that of the artificial light schedule to which it had most recently become adjusted. In the example described the bird would continue to become active between midnight and one A.M., even under the constant conditions. As with the animals subjected to sudden transportation hundreds of miles east or west, its biological clock keeps the bird on an activity schedule wholly out of phase with the outside world, thus demonstrating that it is not being affected seriously by local environmental factors leaking in some unsuspected manner into the constant-temperature chamber.

The dependence of directional choices on a bird's internal time-keeping mechanism has been demonstrated by experiments of this sort, in which the biological clocks of birds are reset by methods essentially like those described. Starlings that Kramer and his colleagues had trained to seek food from one of several feeders around the edge of a circular cage were exposed to artificial light schedules that reset their clocks. A starling that had previously learned to seek food to the east by sun-compass orientation under the natural day-night cycle, was reset so that its artificial "dawn" came six hours later than the natural sunrise. When placed in the experimental cage it now looked for food in the wrong direction from the sun. If, for example, the test was made at the real noon (shortly after daybreak on the bird's artificial cycle), the starling sought food from the feeder directly toward the sun, instead of the one ninety degrees to the left. The latter, of course, was the actual east feeder. The same starling had accurately chosen the east feeder when tested in the same apparatus before its clock was reset. The possibility of resetting biological clocks, by exposing birds to artificially shifted cycles of light and darkness, provides experimenters with a method for determining the degree to which sun-compass orientation is being employed in various situations. The next chapter will consider a variety of experiments that demonstrate how the sun and other celestial bodies are used by birds, in conjunction with biological clocks, to choose appropriate directions both in daylight and at night.

Chapter 8

CELESTIAL NAVIGATION

Gustav Kramer not only initiated the experiments on sun-compass orientation described in the last chapter, but also studied the orientation of nocturnal migrants. He chose European warblers for experiments in a circular orientation cage similar to the one used with starlings in daylight. The top of the cage was exposed to the night sky, and the birds were observed closely for many hours in the season when they were exhibiting migration restlessness. Since Kramer wished the bird's view of the sky to be as nearly the same as during actual nocturnal migratory flights, he used no artificial illumination in these experiments. Instead, he watched the warblers through a transparent plastic floor of the orientation cage, while lying on his back and looking up at the bird silhouetted against the stars. The cage was surrounded with an opaque cylindrical wall to shield the birds from lights and prevent them from seeing any local landmarks. Perches were provided at the center and also around the edges of the cage. The cage walls and roof were constructed from fish netting of the appropriate-size mesh to prevent the birds from colliding violently with any hard surface.

A necessary preliminary to these experiments was the hand-rearing of fledglings. Uncounted hundreds of insects had to be collected for food, and months of patient loving care were required before a bird was fully grown and ready for testing in the orientation cage. Even when this type of experimentation was in full operation, a staff of three or four people could care for only about a half-dozen warblers in any one season, and of these half dozen, only one or two gave clear-cut headings.

A few individual warblers would show clear indications of preparation for flight, or even the preliminary motions of flight itself, on many nights in the main season of migration. These preliminaries were examples of so-called intention movements, small-scale versions of some vigorous activity that often follows them under natural conditions. Sometimes the restless, caged migrants fluttered their wings while still grasping the perch. At other times they would fly back and forth from the perch on one side of the cage to the opposite wall. There, a warbler might flutter briefly against the cloth netting, or rest again on the perch near the side wall toward which it had just flown. On other occasions the bird would fly a few inches upward and hover for a few seconds, facing in one direction.

One warbler that fluttered back and forth for prolonged periods, and yet refrained from dashing wildly against the cage walls, was selected as a suitable subject for more concentrated attention. Kramer noted how much time this bird headed in various directions while fluttering or otherwise indicating an apparent intention to fly off. Sometimes there would be no activity or very little trend to the headings, but on other nights the bird would exhibit clear and consistent choices of certain directions far more often than others. It was impracticable to note to the nearest degree every

momentary shift in the position of a nervously active
bird. Hence, Kramer's usual method was to record
how many seconds the bird spent heading in various
directional sectors, counting only time when it was
fluttering or otherwise giving recognizable indications of directed migration restlessness.

Orientation of Migration Restlessness

The warblers tended to head in directions roughly
the same as those in which they would have been
migrating, had they been at large under natural conditions. Furthermore, the directions of their
headings in the orientation cage changed with the
season, roughly south in fall and north in spring. In
between seasons of migration there was little restlessness in the cages, and no significant directional
tendency. This result was interesting enough, but
in the course of experiments under various weather
conditions it turned out that overcast skies disrupted
the headings altogether. When the stars were not
visible, the headings scattered widely, and were
either quite random, or tended to concentrate
slightly in the direction of any slightly brighter portion of the sky. If the experiments were conducted
near cities, the artificial lights reflected from the
clouds seemed to influence the birds' headings.
Similarly, if the moon was visible the birds often
headed in its direction. This sort of heading toward
the brightest available light in an otherwise nearly
dark cage does not seem to be part of the normal
process of orientation during nocturnal migration.
Perhaps it is more closely related to a bird's tendency
to escape from its darkened cage out into brighter
light. Another possibility is that without stars for
guidance birds tend to fly toward any bright light.

Possibly this explains why under certain conditions they are attracted to lighthouses and "ceilometer" searchlights at airports—often with disastrous results.

The Sauers' Study of Bird Song and Orientation

A few years after Kramer's discovery that migration restlessness was sometimes correctly oriented, similar experiments were started by Franz Sauer, an energetic young zoologist at the University of Freiburg. Sauer and his wife had been engaged in another, equally laborious type of research on bird behavior, the study of how birds come to sing the particular patterns of song that are characteristic of their species. It had been a matter of active debate whether young songbirds learn from their parents, or other adults of their species, just what sequence and pattern of notes is appropriate for a male to employ in announcing that he has staked out a territory and is ready to set up housekeeping. A direct, but burdensome approach was to raise young birds from the egg in complete isolation, in soundproof rooms, where they could grow to maturity without any chance to hear the songs of other birds. The Sauers chose European warblers for this purpose, because they have characteristic songs that differ from one closely related species to another. But raising these insectivorous birds in isolation required the same tireless devotion Kramer had to give to his warblers. In the course of their experiments on the development of singing ability, the Sauers necessarily became expert at this art of mothering nervous and delicate baby warblers until they grew to healthy adult birds ready to breed, sing, or migrate. The outcome of the experiments on song development

was fairly clear evidence that in European warblers
an almost normal song is produced even in total
seclusion with no possibility of learning by hearing
other, experienced birds. This topic is outside the
scope of this book, but interested readers will find
it fully and lucidly discussed in W. H. Thorpe's book
Bird Song (Cambridge University Press, 1961).

The Sauers used a circular cage similar to Kramer's,
and they studied the headings of their birds not only
under the natural sky but also in a planetarium. This
"artificial sky" had many advantages for experiments.
Cloudy weather was no problem, and the heavens
could be imitated not only as they would appear
naturally at Freiburg on the night of the experiment
but as they would appear at some other place on the
earth's surface, or at some other season. Furthermore, the Sauers had available adult warblers, in
full breeding condition, but totally inexperienced
with the real world outside their soundproof rooms,
birds that had never seen the sky at all.

The results of the Sauers' early experiments were
as dramatic and exciting as Rowan's first demonstration that lengthened days in winter would bring
birds prematurely into breeding condition. Certain
especially co-operative warblers in the Freiburg
laboratory did indeed show consistent directional
tendencies that corresponded to the normal migratory headings of their species at the season of the
experiments. As illustrated in Figs. 21 and 22, these
headings were equally accurate whether the birds
were shown the natural sky or the star pattern reproduced on the dome of a planetarium. As in Kramer's
experiments, cloudy skies elicited only disoriented
and virtually random choices. Evidently these warblers were indeed able to select the appropriate direction for their migration on the basis of the stars.

Furthermore, these directional choices were reversed with the season. The same bird exposed to approximately the same star pattern, in the same experimental situation, would head north in spring but south in fall. By injections of appropriate hormones the Sauers were even able to bring one or two warblers into breeding condition in between the normal seasons of migration, and these birds showed the spring headings.

All this was remarkable enough, but the Sauers' experiments also indicated that in at least two individual warblers the appropriate directional response to the star pattern occurred the first time the individual bird was allowed to see the sky. These experiments, summarized in Fig. 22, were carried out with hand-reared warblers that had been kept indoors ever since they were very young nestlings. Yet when placed in the orientation cage as adults, in periods of migration restlessness, they made essentially the appropriate choices of direction. This

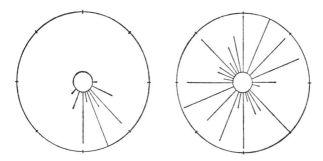

Fig. 21. *Directional headings of a warbler during the autumn period of migration restlessness. The orientation cage was in a planetarium. The length of each line represents the proportion of the time that the restless bird spent heading in the direction indicated. At the left, headings under the autumn sky of Germany; at right, random headings with diffuse illumination from the planetarium dome.*

result is so striking that many biologists have found it difficult to believe. The natural next step of repeating the experiment has been very difficult, simply because of the enormous labor involved in hand-raising birds such as European warblers and keeping them sufficiently healthy to show migration restlessness and give headings in an orientation cage.

The implications of the Sauers' experiments are far-reaching, because they apparently show that, built into the organization of the brains of migratory birds, is some mechanism that causes them to react in a specific way to the pattern of stars. Such a mechanism must also include provision for a reversal of this directional choice between fall and spring, presumably dependent on the bird's internal physiological and endocrine balance. Presumably the genetic make-up of the particular species must determine whether or not such directional choices are made, and to some degree must dictate the angle relative to the star patterns that the bird will select.

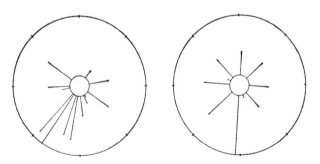

Fig. 22. *Directional headings of two warblers reared in complete isolation and then tested in the Sauers' orientation cage under the outdoor sky in late summer or early fall. Both birds tended to head south to southeast during their migration restlessness, even though they had never before seen the sky or star patterns on which this orientation must have been based.*

If the Sauers' findings apply generally to migratory birds, then various species must have genetically determined recognition patterns appropriate for the latitudes and seasons of their migrations. Penguins swimming through the South Atlantic have very different star patterns to guide their migrations from those that are visible to plovers starting south from the shores of the Arctic Ocean.

Hamilton's Experiments with Bobolinks

Recent experiments by William J. Hamilton, III, in San Francisco, have shown that bobolinks also make directional choices during migration restlessness, when they have only the stars as directional cues. This species performs a very long migration from well south of the equator into the northern United States and southern Canada. Hamilton found bobolinks somewhat more variable in their directional responses in orientation cages than the Sauers' warblers had been in the most clear-cut of the latter's experiments. To the Sauers' procedures Hamilton added the playing of tape recordings of bobolink call notes, a conspicuous feature of the nocturnal migratory flights of bobolinks and some other night migrants. The recorded call notes served to intensify the directed intention movements of the bobolinks in the orientation cages. One complication that developed in Hamilton's results with bobolinks, to a greater extent than in the Sauers' earlier experiments with European warblers, was a bird's tendency to alternate between southward headings in the fall and choices of the exact opposite direction. Even when these south and north headings were alternating in an unexplained fashion, some sort of celestial orientation was evident, since intermediate, easterly or westerly headings were quite rare. Two bobolinks that had been hand-reared from a very early stage of development showed directional choices of this type. But more than birds captured as adults would do, they displayed a tendency to oscillate back and forth from the appropriate southward heading in autumn to a reversed, northward choice. Even these bobolinks that were seeing the star patterns for the first time were not random in their headings, for north and south predominated heavily over east or west. Hamilton's experiments with bobolinks provide at least partial confirmation of the Sauers' remarkable finding that directional reactions to the stars are genetically built into the nervous systems of migratory birds.

Even in the clearest of the Sauers' or Hamilton's experiments in orientation cages the majority of the headings are spread over forty-five to sixty degrees, with a few stragglers scattered in many other directions. This dispersion requires that any difference between headings of two groups of birds must be distinctly greater than sixty degrees before it can be accepted as significant. In some of the Sauers' planetarium experiments warblers that had shown directed migration restlessness under an accurate reproduction of the autumn sky over Germany were then presented with the sky as it would appear at latitudes farther to the south. It seemed in some cases that the headings turned from southeast to south at just the latitude of the Mediterranean Sea, but the overlap between the two sets of directional choices was not sufficient to make the evidence firmly conclusive. Pending more extensive and detailed experiments of this type, most students of the matter are convinced only of the Sauers' basic finding that northward or southward headings are displayed under the real or planetarium sky at the appropriate seasons. Further embellishment upon the available data does not seem warranted.

Directional versus Goal-directed Orientation

It is important to distinguish two distinct types of orientation that seem to be employed by birds in natural migration, in homing experiments, or in orientation cages. The simplest is often called *directional orientation*. This term means that the bird has the ability to select a particular compass direction and, presumably, to fly in that direction if free to do so. Kramer's or the Sauers' migrants which headed roughly south in fall would be good examples. The second type of orientation is called *goal-directed orientation* and is exemplified by Matthews' experiments with homing pigeons and Manx shearwaters described in Chapter 7. These birds flew off approximately toward home regardless of the direction in which they had been carried from their loft. Goal-directed orientation is obviously more difficult to explain than directional orientation, because the bird must somehow determine the appropriate direction to reach its home or other goal area, and then manage to select that direction. Furthermore, goal-directed orientation is more difficult for an experimenter to demonstrate. No single release point can suffice for such a demonstration, no matter how many birds are employed, nor how accurately they head toward home. Only if they head for home from two or more different directions can we conclude that they are exhibiting goal-directed orientation.

Kramer put the key question rather well when he pointed out that for birds sun-compass or star-com-

pass orientation literally takes the place of a man's
use of a compass to find his way. But a compass is
of little use to man or bird unless he knows the direction in which he wishes to travel. In ordinary human practice we use compasses (or the equivalent)
in conjunction with maps, or some knowledge of
geographical relationships that serve the same purpose as a map. Before his untimely death in a
mountain-climbing accident, Kramer emphasized on
many occasions that biologists could account for the
compass used by homing birds, but not the map. In
other words, he felt that some other, and still completely unknown, factor must be combined with sun-
compass orientation to explain the homing of the
best strains of pigeons, as well as the navigation of
Manx shearwaters.

The Sun-Arc Hypothesis

When Matthews was faced with these same facts,
he attempted to explain them by assuming that the
sun alone suffices to provide birds with the information needed to achieve essentially correct homeward headings and display a high order of goal-
directed orientation. As explained in the last chapter,
sun-compass orientation implies the use of the birds'
biological clocks as a basis for selecting a progressively changing angle relative to the sun's position
in the sky. This ability has been amply demonstrated
not only in birds but also in lizards, fishes, insects,
and in many other invertebrate animals. Expressed
in human terms, one might imagine that the birds
say to themselves, "It is now noon, and since I want
to go north I should fly away from the sun," or later
in the day, "Now it is fairly late in the afternoon;

therefore, to fly north I must keep the sun over my left shoulder."

Matthews postulated that the birds might compare the sun's altitude above the horizon with the altitude that could be expected at the bird's home at the time in question. Transposed again into human expression, the bird's thought might be, "I know by my biological clock that it is early morning, but the sun is too high for this time of day; either they have carried me south, where the sun is higher in the sky at this hour, or they have carried me east, where it rose earlier than back home." At this point one might well stop imagining bird geometry and assume that our bird, having guessed this much, would decide to make the best of an uncertain situation, in which its location could be anywhere between east and south. This it could do by splitting the difference and heading northwest.

An English ornithologist, Tunmore, postulated a type of behavioral reaction on the bird's part that can be expressed even more simply. His hypothetical rule for a bird finding itself in unknown territory after a homing experiment is this: "If the sun is higher than you expect at your home area, fly away from the sun. If the sun is lower than expected, fly toward it." Much of the time a bird obeying Tunmore's rule would indeed head within thirty or forty-five degrees of the homeward direction. But at other times of day, especially near sunrise and sunset, Tunmore's rule would produce deviations of ninety degrees or more. Some of the best homing performances and initial headings of Matthews' pigeons occurred under these conditions. But the simplicity of Tunmore's rule is very appealing, and even though it may not apply to all birds under all conditions, perhaps it can explain some of the essentially correct initial headings that have been observed by Matthews and Kramer.

Matthews postulated that birds do something rather more complicated—namely, observe the sun's movement through the sky and, on the basis of its rate of ascent or descent relative to the horizon, extrapolate its arc across the sky to the noon, or highest position. He assumed that the bird judges not only how high the sun would be at noon, but how long a time before or after the moment of observation the sun would in fact be in its noon position. If all this has been accomplished with sufficient accuracy, the bird could then judge the direction of its displacement by the experimenter. Success would depend, of course, not only on the precision of the bird's measurement of the sun's altitude and rate of change of altitude, but also upon the accuracy of its internal clock. Since birds have not been shown to possess any of these measuring abilities to the necessary precision, and because of the complexity of the postulated extrapolation of the sun's arc, Matthews' theory has not won wide acceptance. But a considerable amount of homeward orientation could be explained by some modified form of sun-arc theory, such as Tunmore's simple rule, "Fly away from the sun if it is unexpectedly high, and toward it if it seems too low." Once again a full explanation awaits the initiative of future investigators.

Directional Orientation under Natural Conditions

Directional orientation can be based on the sun or stars, but there are also other ways it can be accomplished, at least in theory. Sea birds might maintain their headings by flying at some definite angle relative to the patterns of waves on the sea below them, although this has never been demonstrated conclusively. If this kind of navigation does occur, it would be an instance of directional orientation based on a consistent visual signal from the environment. Or birds might fly consistently downwind, which, as far as we know, they could only do by visual reference to the ground to determine which way the wind is influencing their flight paths. When evidence becomes available to show that a migrating bird is employing a certain type of directional orientation, we shall obviously have progressed a very long way toward accounting for its ability to maintain a correct course on a long migratory flight. A further problem is to account for changes in direction that may occur at certain points along the way, such as the shift from southeast to south when shore birds from northwestern Canada reach the Gulf of St. Lawrence. There also remains the truly formidable problem of explaining how the nervous system of a bird is organized to recognize a celestial signal for one directional orientation in the spring and another in the autumn.

Genetically Determined Migration

When birds maintain an appropriate direction during their normal migrations, there is one possible explanation that has not yet been considered. This is reliance on a tradition that might be passed on from one generation to the next, each young bird learning the route by flying either with its own parents or with older birds of the same species. Ducks and geese appear to migrate in family flocks, and might be suspected of relying more than other birds on traditional migration routes. Specific experiments have been designed to test this possibility, and one of the earliest was performed by William Rowan a few years after his experiments with altered light schedules that were described in Chapter 3. In these orientation experiments Rowan kept young crows in large outdoor cages until late autumn, when all the wild crows had migrated south from the vicinity of Edmonton, Alberta, to the area of Kansas and Oklahoma. These crows were not subjected to any experimental treatment, except detention in Alberta until mid-November. Public announcements encouraged crow shooting at the time they were released, and Rowan received a number of reports and recoveries of crows that were conspicuous because of the wintery conditions. Most of these inexperienced crows had traveled south or east from the release point, evidently following the customary migration route for population from which they had been drawn. (See Fig. 23.)

A similar, but more extensive, series of experiments has been carried out more recently by Frank C. Bellrose with juvenile blue-winged teal that were caught and banded, along with other migratory ducks, at a large waterfowl refuge in central Illinois. Their normal pattern of fall migration already had been determined by recoveries of banded teal. About 80 per cent of these recoveries, including both juveniles and adults, were between south and southeast of the banding station in Illinois. The most distant recoveries were in Guatemala, Panama, Colombia, Venezuela, Jamaica, Haiti, and Cuba. Other teal had been shot in Louisiana, Alabama, Georgia, and Florida, probably while on their way to the Caribbean.

Blue-winged teal normally leave Illinois by the end of October, and only a few stragglers remain as late as the middle of November. During the four seasons of his experiments, Bellrose held a total of 1271 juvenile blue-winged teal in cages, and released them between November 10 and December 8. These young teal already had traveled at least two or three hundred miles from their nesting areas. A total of 111 of these birds were later recovered at a significant distance, and most of these recoveries were south to southeast of Illinois. Figs. 24 and 25 show absolutely no appreciable difference in the distribution of recoveries whether the birds had been released immediately after banding or detained until all adults of their species had departed. One of these juvenile teal, captured on its first fall migration through Illinois and held until November 10, was shot two days later by a hunter in Florida. Two others were shot in Mobile Bay, Alabama, two days after release in Illinois. These birds had flown at least 350 miles per day without any adults of their species to guide them. Their migratory success could not be accounted for by favorable winds, nor was it likely that they could have joined flocks of other species of ducks, because the only other species passing through Illinois this late in the season migrate in somewhat different directions. In further experiments Bellrose shipped 895 mallard ducks from Illinois to Utah, and those that were shot later the same fall had migrated south from the release points in Utah, rather than compensating for their displacement. Evidently the tendency to fly south takes effect even in totally strange surroundings.

Bellrose's experiments did not present any direct evidence about the means of orientation used by these migrating teal. But William J. Hamilton III trained teal chicks only a few days old to find water (which they needed urgently because he provided only dry food) by searching in one specific direction

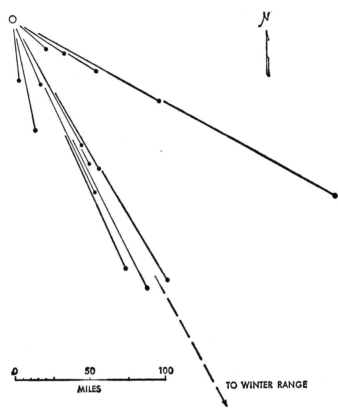

Fig. 23. Rowan's experiment with young crows detained in outdoor cages near Edmonton, Alberta, until late November, some weeks after all wild crows had migrated about 1500 miles southeast to Kansas and Oklahoma. Each dot represents the place where one crow was shot a few days after release.

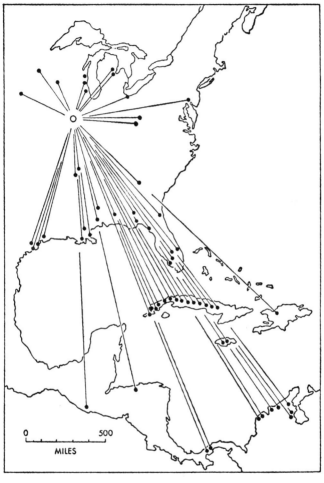

Fig. 24. Recoveries of blue-winged teal released immediately
after banding in early fall as they migrated through Illinois.

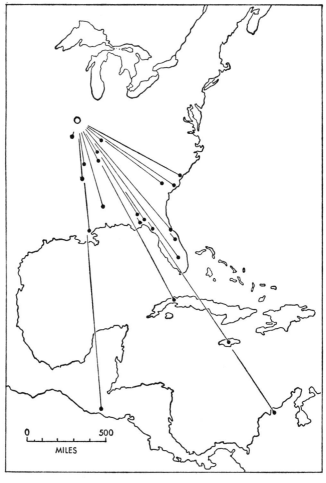

Fig. 25. Recoveries of young blue-winged teal held in Illinois until November and December, long after all others of their species had migrated south.

in an orientation cage similar to Kramer's or the Sauers'. This cage allowed a clear view of the sun, but no local landmark was visible to the ducklings. Hamilton's downy young ducks, like Kramer's starlings or Sauer's warblers, learned to select the direction that would lead them to water, but only when the sun was visible. If it was hidden by clouds, they were quite disoriented. The same orientation was obtained also at night, provided the stars were visible. The fact that the same species can learn to make directional choices on the basis of the sun and stars suggests that celestial cues were used by Bellrose's young mallards flying south from Illinois.

The white storks that nest in Europe also have been used for experiments in which young birds were held in captivity on the fall migration until all adults had left for Africa. Storks nesting in east of central Germany migrate southeast in fall, and detour around the Mediterranean through Syria and Egypt, as indicated in Fig. 8. Storks nesting in the Rhine watershed, France, and Spain cross the Staits of Gilbraltar and then spread out through Africa. Young storks from Eastern Germany were raised in the Western Zone, where the local storks migrate toward Gibraltar, and when these birds were released at the normal season of fall migration they also tended to migrate southwest. But they may simply have accompanied storks native to the western area. Another group from the Baltic coast of East Prussia was held in captivity until after the storks of the Rhine Valley had flown southwest in the fall. They were released at Essen, on the Ruhr River, a tributary of the Rhine. Most of these storks flew south or southeast, rather than southwest, the trend suggesting that they had a genetically determined tendency to fly in the customary direction for the eastern population when they were not induced to accompany other storks native to the area of release.

The German ornithologist Werner Rüppell studied a migratory population of hooded crows that could be trapped and banded in large numbers during both spring and fall migration as they passed through a large bird-banding station on the Baltic Coast of East Prussia. Recoveries of crows released immediately after banding showed that their normal winter range was central Germany, the southern Netherlands, and Belgium, but that they did not reach the North Sea coast of Germany. In summer they migrated to an area northeast of the banding station that included Lithuania, Latvia, Estonia, Finland, and adjacent parts of Russia. A large number of these crows, caught on the shores of the Baltic on their spring migration, were transported west to be released at Flensburg, just south of the Danish border, and well outside their winter range. The subsequent recoveries the following spring and summer were in Denmark and Sweden, indicating that the crows had continued to migrate north or northeast just as they would normally have done east of the Baltic.

A Dutch ornithologist, A. C. Perdeck, carried out a similar experiment with starlings that he was able to capture in very large numbers on their fall migration through The Hague. This population of starlings had been shown by recoveries of banded birds to spend the summer near the Baltic coast of Denmark, northern Germany, Poland, and eastern parts of the Netherlands. After passing through The Hague, they moved generally westward to spend the winter in the western parts of the Netherlands, Belgium, northern France, southern England, and Ireland. Perdeck transported more than 11,000 starlings south from The Hague to release points in Switzerland immediately after they were captured in September, at the beginning of the fall migration. Several were recovered later the same autumn, and the distribution of the recovery points depended on the birds' ages. The young starlings that had been carried from Holland to Switzerland at the time of their first fall migration were recovered in southern France, Spain, and Portugal. They were continuing apparently in the same westerly direction in which they had already flown some distance. The adult starlings, on the other hand, tended to travel northwest into the region where they would normally have spent the winter. It is not clear what guided these older starlings back to their usual winter range. Since enough time had elapsed to have permitted flight over many times the distance between the points of release and recovery, they may have searched a wide area before finding familiar winter territory. But Perdeck's experiment suggests that the older starlings used some form of goal-directed orientation. In any case the young starlings were exhibiting a relatively simple directional orientation.

Puzzling Directional Orientation

Shortly after Matthews' and Kramer's experiments had been reported, T. H. Goldsmith and I decided to study the initial headings of common terns, which nest in considerable numbers on Penikese Island near Cape Cod. We selected suitable inland release points with a clear view in all directions, and as free of lakes that might prove distracting as any place we could find in southern New England. Wishing to keep terns in captivity as briefly as possible, we first released them at Storrs, Connecticut, the closest suitable release point that seemed far enough inland so that terns would be unlikely to see the ocean. To our pleased surprise, the terns did take off quite consistently southeast, and the average of their initial headings was almost directly toward their home island. Furthermore, when we repeated these experiments at other inland release points the same southeast headings were evident.

But all our release points were roughly northwest of Penikese Island, and, as pointed out previously, accurate homeward headings from a single direction cannot be accepted as goal-directed orientation. We therefore brought additional terns from a nesting colony on the coast of Maine to the same release point at Storrs. For these birds the home direction was northeast. To our further surprise, and disappointment, the Maine terns headed southeast just as consistently flying toward Penikese Island as had the terns that nested there (see Fig. 26). Neither group showed any consistent headings at all under overcast skies. Hence, we were dealing with a type of directional sun-compass orientation. It is not at all clear why common terns released inland should fly southeast, although one theory is that such a habit helps them to regain the Atlantic coast if they are accidentally carried inland by storms. But we later found that common terns nesting near Lake Michigan also headed southeast, as did control terns from Massachusetts, when both were released in upstate New York.

An equally puzzling case of directional orientation has been discovered in mallard ducks. Matthews has studied a nonmigratory population of these very abundant ducks at the Slimbridge Waterfowl Research Station on the Severn Estuary near the west coast of England. Initially he wished to determine whether they would home to Slimbridge when released in flat, open country away from water courses. To his surprise, they always tended to fly northwest, regardless of the direction in which they had been carried away from Slimbridge. Bellrose carried out

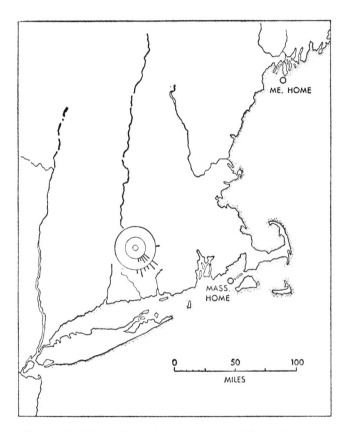

Fig. 26. Initial headings of terns transported from their nests on two islands off the coasts of Maine and Massachusetts to the same inland release point in Connecticut (small circle at center of directional heading diagram). The inner bars represent birds from the Massachusetts colony, the outer bars terns from the island in Maine. Each short bar represents a single bird; the longer bars show directions taken by two birds.

similar experiments with mallards in central Illinois, and his ducks tended to head north, even though they had been trapped on their fall migration. Both in England and in Illinois the mallards scattered at random under overcast skies. Like our New England terns that headed southeast, mallard ducks showed their northward headings only if the sky was clear.

"Nonsense Orientation"

In all three cases sun-compass orientation was underway, but the particular choice of direction has so far defied explanation. Matthews has shown recently that another population of mallards, from the parks of London, tends to fly off in other directions than the northwest headings of their Slimbridge relatives. Matthews calls this type of directional orientation "nonsense orientation," implying not that it serves no biological purpose whatsoever, but that its function is unknown to us. Many of the mallards that Matthews observed heading northwest were later recovered in all directions around the release point. Evidently the consistent northwest headings are maintained for only a short time. Possibly this behavior makes it more likely that a flock dispersed by danger can reassemble shortly afterwards.

The nonsense orientation of Matthews' mallards seems rather uninteresting in comparison with the star navigation of the Sauers' warblers, or the transatlantic homing of a Manx shearwater. But mallard ducks have proved to be excellent experimental animals, because they are abundant, live well in captivity, and because their northwest headings are so consistent, even though as yet they make no sense to us. Matthews has used this aspect of mallard behavior to discover some significant new facts about the celestial navigation of birds. He released many groups of mallards at various suitable points in all directions from Slimbridge, at all hours of the day, under clear skies, and when the sun was completely hidden by clouds. Neither the location of the release point relative to Slimbridge nor the time of day made any appreciable difference in the consistent northwest headings. Aside from very strong winds, which deflected the birds to some extent, the weather was without effect so long as the sun was visible, either directly or as a recognizable bright spot in a thin layer of clouds.

Both Matthews and Bellrose have released mallard ducks at night. In order to follow them in the dark they attached small flashlights to the ducks' legs. These lights consist of the battery from a "Penlite" and a bulb, fastened to the battery terminals just before attachment to the bird. The lights can be followed at night for about as far as a flying duck can be seen through binoculars in daylight. The battery is secured to the duck's leg by paper tape that softens quickly in water; the bird does not remain encumbered even with this minor burden for more than an hour or two after it reaches some pond or stream. On one occasion when I had the opportunity to accompany Matthews on a release during the night, one of the ducks chose to land in a deep roadside ditch instead of flying off immediately. The flashlight traced its underwater course as the mallard dove and swam for a few minutes until the tape soaked through and the light fell off.

The results of many observations of initial headings at night were similar to those made by day. If the skies were clear, the Slimbridge mallards tended to fly northwest. Bellrose also found that Illinois mallards headed north both in daylight and at night.

Evidently these ducks can use the stars as well as the sun, and their nonsense orientation is equally consistent whether based on sun-compass or star-compass orientation. Matthews observed on a few occasions that his mallards flew northwest when the stars were obscured by a thin cloud layer but the moon remained visible. Hence "moon-compass orientation" is also practiced by mallards and presumably by other birds as well.

North Star Mallards

Matthews next shifted the biological clocks that must have been operating in the ducks to permit them to head northwest consistently, regardless of the time of day or the sun's apparent position in the sky. As we have seen, Kramer and others had discovered that when birds were displaying sun-compass orientation in circular cages, they made the expected directional errors after their internal "clocks" had been reset. Matthews used exactly similar methods to reset his mallards' clocks. He kept them in closed rooms under an artificial schedule of light and darkness for several days, until their cycles of activity, feeding, and sleeping were altered to conform to the new conditions. One group of ducks was shifted six hours ahead, another six hours behind the actual time of daybreak and dusk, and a third group was shifted twelve hours out of phase with the true day and night. Finally a control group was kept in outdoor cages under completely natural daylight and darkness.

When these four groups of mallards were set free on clear days, Matthews found the expected changes in their headings as a result of the resetting of their clocks. The controls headed northwest as usual.

Those whose clocks were set ahead six hours behaved as though it was six hours later in the day, and headed roughly southwest. Similarly the ducks whose artificial schedule was six hours slow deviated in the opposite direction. The mallards twelve hours out of phase with real time headed southeast, just opposite to their usual northwest headings. As one can see from Fig. 27, all these directional choices show a variation of about plus or minus thirty degrees, but the effects of resetting the birds' clocks are so clear that there can be no doubt as to their reality or significance.

These results obtained under sunny skies were in complete accord with expectations based on the previous discoveries of Kramer and his colleagues. But there was an unexpected difference when Matthews released his mallards at night, under clear skies, after resetting their biological clocks in exactly the same fashion, six hours fast, six hours slow, and twelve hours out of phase. *Under the stars all three groups showed the same northwest headings,* whether or not their clocks had been reset. This experiment tells us at once that the birds can learn more from the stars than from the sun. The sun provides only a single point of reference, and with the aid of their internal clocks birds and other animals can use it for a directional, sun-compass orientation. But Matthews' mallard ducks were using the stars for a new and more impressive type of directional orientation, which was not thrown into error, as is sun-compass orientation, by resetting their internal timing mechanisms.

There has not yet been time for Matthews' latest discovery to be followed up by other investigations that may help to explain just how the mallards use the pattern of stars to select their favored northwest

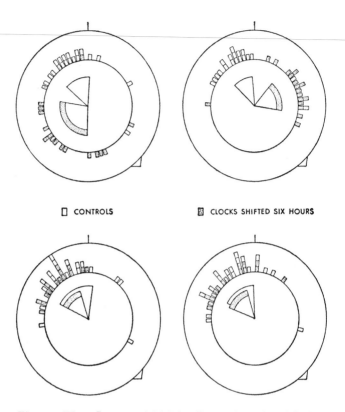

CONTROLS ☐   CLOCKS SHIFTED SIX HOURS ▨

Fig. 27. The effects on initial headings when the biological clocks of mallard ducks are shifted. In all diagrams the open rectangles represent the headings of control birds that had not been subjected by Matthews to any clock-shifting treatment; these ducks tended to fly northwest. Stippled rectangles show headings of birds kept for several days on a shifted light cycle that reset their biological clocks six hours ahead (left-hand graphs) or behind (right-hand graphs). The sectors at the center of each graph show the angular range within which half the headings were contained. The two upper graphs show results obtained in daytime when birds had a clear view of the sun; the resetting of biological clocks produced the expected changes in headings. Evidently these birds were using sun-compass orientation.

The two lower diagrams show the result of the same experiment performed on clear but moonless nights. With the full star pattern available, the directional headings were not significantly affected by resetting the biological clocks. Were these ducks using the North Star?

direction. One obvious possibility is that they rely on the North Star, or some group of its immediate neighbors. If so, they must have learned that these particular stars are reliable guides for orientation. Even when they find the rest of the heavens ninety degrees from their expected positions, the ducks ignore this discrepancy and fly off somewhat to the left of the North Star. Matthews may be able to obtain evidence on this point if, in the future, he is able to release mallards with their clocks shifted when a large cloud covers the sky around the North Star but leaves other stars visible. Reliance on the North Star seems to us the simplest way to achieve star-compass orientation in the face of uncertainty or error in one's judgment of time. But it is not the only aspect of the star pattern that would permit correct star-compass orientation, at least in theory. Since additional investigations will clearly be required to settle this question, it would be premature to discuss it further here.

*   *   *   *   *

After the past twenty years of research on bird navigation it is clear that migrating birds are quite capable of setting courses by the sun, moon, or stars. Before 1940 it seemed absurd to think that they might correct for the apparent movement of sun or stars across the sky and use celestial guides to the direction appropriate for annual migrations. Still more ridiculous was the notion that the North Star might be picked out from all its fellows as a beacon for migratory flight. This last hypothesis has not been established, but it is now at least a reasonable one, well worth testing. If it is not correct, the birds in Matthews' latest experiments must be accomplishing some equivalent feat of star mapping.

Occam's razor and the "molecular" approach proved in this case to be unduly limiting. The actual facts turned out to support what had seemed romantic speculation. This does not mean that wishful thinking will prove correct in more than a handful of other cases. But the celestial navigation of migratory birds has demonstrated that all facets of biology are not yet reduced to dull and routine affairs. So many biological phenomena still defy adequate explanation that we may realistically hope, in due course, for other future developments as pleasantly surprising.

NEXT STEPS

While biologists have learned a great deal about bird migration, the preceding chapters should also make it clear that for years to come many unsolved problems will remain to fascinate curious scientists of all ages. What are our principal areas of ignorance about bird migration and what, in the foreseeable future, can be done about it?

It would obviously be helpful to know more about actual routes flown by individual migrating birds. If we had a number of accurate maps showing just where particular birds had flown on long migrations, we might be able to identify important factors influencing the timing and orientation of these flights. For example, if abrupt turns are made at a certain point in space and time, they would pinpoint important parts of the whole pattern of migratory behavior deserving closer scrutiny. Keeping track of migrating birds by visual observation from airplanes has not appealed to many other investigators since my studies of homing birds fifteen to twenty years ago. This method is limited to daylight hours and to large and conspicuously colored species of birds.

Recent developments in the technology of transistors and similar devices have now made it possible to build radio transmitters small enough to be carried with reasonable comfort by large and medium-sized birds. Glamorous press releases depict the tracing of barn swallows by radio receivers from North to South America, but in fact the limitations of this method are still severe. For the next few years radio telemetry of migrations seems likely to be limited to birds of the size of pigeons or larger. The experimenter must trade off range of detection and battery life, on the one hand, against weight and size of antenna, on the other. Last but not least, the greater the miniaturization, the more costly are the components required. Short wave lengths must be used in any designs that seem even remotely practicable in present circumstances, and consequently the transmission pathways are virtually limited to straight lines. Birds will not be detectable over distances such that the curvature of the earth intrudes itself between transmitter and receiver. An airplane a few thousand feet in the air has obvious advantages in this regard over receiving stations on the ground.

Radar also will provide more information about bird migrations, but until radar systems have been specifically designed for bird watching (or at least operated so as to maximize bird echoes rather than minimize them in favor of echoes from airplanes or missiles), radar data will probably be limited to the densities and directions of mass flights. Ideally, one should be able to trace individual birds, and identify them at least roughly by determination of their size and the rate at which their wings are beating. Perhaps in due course it will be possible to employ airborne radar systems that can scrutinize migrating birds while circling around them, much as we once watched homing gannets with our unaided eyes.

It should not be forgotten that bats also migrate for hundreds of miles, and that their vision is so poor that celestial navigation seems almost out of the question. Far too little is known about the timing of bat migrations, the directions flown, and whether flights occur under overcast skies or only when moon or stars are visible. These and other basic questions require answers before one can even formulate theories of bat navigation, let alone test them. The high-frequency sonar signals that bats use for avoiding obstacles and catching insects are absorbed almost totally in traveling a few hundred feet through the air, so that long-range navigation by echo seems even less likely than navigation by the moon. But this entire subject will remain in the realm of speculation until someone learns how to experiment directly with the orientation of migratory bats.

Experiments with orientation cages, such as those used with such success by Kramer and the Sauers, unquestionably will tell us still more about the factors that influence directional choices of birds during migration restlessness. Such experiments probably will be improved as biologists learn more about the technique of adapting nervous song birds to confinement. Possibly a modified wind tunnel could be used to allow the experimental birds to take wing and perform something much closer to true migratory behavior while choosing their migratory headings. Particularly important experiments are needed to analyze as thoroughly as possible the inherited tendency for young birds to choose a certain direction relative to the star patterns of the night sky. This is such a remarkable discovery that it calls for very detailed checking and determination of its accuracy and consistency. Here again, a limiting factor in such experiments is the down-to-earth know-how required to rear the nestlings and keep them in sufficiently good health and spirits to attempt directed migratory flights under experimental conditions.

In some of the Sauers' experiments that were not described in Chapter 8, warblers in the planetarium were shown the star patterns that were visible at some distant place on the surface of the earth. One bird seems to choose the direction that would have carried it back to its home locality, but the data were too scanty and too variable to be convincing. Nevertheless, the suggestion that the stars can be used for orientation toward a specific goal, as well as for orientation in a general compass direction certainly should be followed up. Until much more work is done with orientation cages in planetaria, the question will remain unsettled.

In daytime, birds seem able to use only the sun for celestial orientation, and by day only goal-directed orientation has been demonstrated in orientation cages. Yet actual initial headings of birds released during the day in Matthews' and Kramer's homing experiments showed reasonably consistent goal-directed (that is, homeward) headings. It is theoretically difficult to account for goal-directed orientation based on a single celestial beacon such as the sun. The bird would obtain much more information if it could see two rather than one celestial beacon, the position of which was dependent upon its displacement in a homing experiment. When the moon is visible by day it might conceivably provide the second point of reference. Alternately, if the bird could resolve even the slightest pattern on the sun's surface this too would provide more information than a homogeneous disk. Unfortunately for this type of explanation, however, accurate homeward headings have occurred when the moon was not visible, and sunspots seem almost impossible to resolve without special filters and some magnification. Could the pecten described in Chapter 4 constitute such a filter? Probably not, since it is a blind spot rather than a screen interposed between the sensitive retina and the light quanta arriving through the pupil.

Deeper into the crystal ball I cannot gaze with enough enthusiasm to inflict further speculations on the patient reader. Attention should be called to only one more point. Real advances in understanding a subject like bird migration almost always come as partial or complete surprises. A generation ago neither I nor any other scientist would have predicted with the slightest confidence several of the discoveries about bird migration summarized in Chapters 7 and 8. Future advances may be equally unexpected. If scientific progress were predictable, it would become a sort of engineering, useful perhaps, but much less fun.

FURTHER READING

Animal Orientation and Navigation. A Symposium edited by S. R. Galler, K. Schmidt-Koenig, G. J. Jacobs, and R. E. Belleville. National Aeronautics and Space Administration Publication No. SP-262. Washington, D.C.: U. S. Government Printing Office, 1972. A series of technical papers by active investigators, more than a dozen of which are directly pertinent to bird migration.

Avian Biology. New York: Academic Press, 1971–74. Three volumes of essays by numerous technical experts. Especially significant is the review of bird navigation by Emlen in Vol. 3.

Dorst, J., *The Migrations of Birds.* Boston: Houghton Mifflin Co., 1962. A thorough account of migration routes and their biological diversity, including many details not appropriate for this small book.

Eastwood, E., *Radar Ornithology.* London: Methuen and Co., 1967. A concise and lucid explanation of how radar has been used to study birds, together with an excellent summary of what has been discovered by this new technique.

Frisch, Karl von, *Bees, Their Vision, Chemical Senses, and Language.* Ithaca, N.Y.: Cornell University Press, revised edition 1971. A classic among studies of animal behavior. It is readily understandable and exemplifies research in this field at its best.

Griffin, D. R., *Listening in the Dark.* New Haven, Conn.: Yale University Press, 1958: Reprinted by

Dover Publications, New York, 1974. A detailed review of echolocation as practiced by bats, birds, and men.

Keeton, W. T., "The orientational and navigational basis of homing in birds." A chapter in *Advances in the Study of Behavior*, Vol. 5. New York: Academic Press, 1974.

Lockley, R. M., *Animal Navigation*. New York: Hart Publishing Co., 1967. A general but authoritative review of major examples of animal navigation.

Matthews, G. V. T., *Bird Navigation*. London: Cambridge University Press, second edition 1968. A comprehensive analysis of experiments on homing and navigation in birds, with special emphasis on the author's theory of sun-arc navigation.

Moreau, R. E., *The Palaearctic-African Bird Migration Systems*. New York: Academic Press, 1972. A scholarly and detailed treatment of the many species which migrate between Africa and the Eurasian land mass, with the maximum emphasis on ecology that available evidence allows.

Orr, R. T., *Animals in Migration*. New York: Macmillan Co., 1970. A broad survey of migration in a wide variety of animals, with brief discussions of ecological and physiological aspects.

Storer, J. H., *The Flight of Birds Analyzed Through Slow-Motion Photography*. Bloomfield Hills, Mich.: Cranbrook Institute of Science, 1948. An elementary discussion of the aerodynamics of bird flight illustrated by numerous photographs.

Welty, J. C., *The Life of Birds*. Philadelphia: W. B. Saunders Co., 1962. A comprehensive and well organized textbook of ornithology, including accurate summaries of migration and orientation.

INDEX

A CATALOGUE OF SELECTED DOVER BOOKS
IN ALL FIELDS OF INTEREST

A CATALOGUE OF SELECTED DOVER BOOKS
IN ALL FIELDS OF INTEREST

AMERICA'S OLD MASTERS, James T. Flexner. Four men emerged unexpectedly from provincial 18th century America to leadership in European art: Benjamin West, J. S. Copley, C. R. Peale, Gilbert Stuart. Brilliant coverage of lives and contributions. Revised, 1967 edition. 69 plates. 365pp. of text.

21806-6 Paperbound $3.00

FIRST FLOWERS OF OUR WILDERNESS: AMERICAN PAINTING, THE COLONIAL PERIOD, James T. Flexner. Painters, and regional painting traditions from earliest Colonial times up to the emergence of Copley, West and Peale Sr., Foster, Gustavus Hesselius, Feke, John Smibert and many anonymous painters in the primitive manner. Engaging presentation, with 162 illustrations. xxii + 368pp.

22180-6 Paperbound $3.50

THE LIGHT OF DISTANT SKIES: AMERICAN PAINTING, 1760-1835, James T. Flexner. The great generation of early American painters goes to Europe to learn and to teach: West, Copley, Gilbert Stuart and others. Allston, Trumbull, Morse; also contemporary American painters—primitives, derivatives, academics—who remained in America. 102 illustrations. xiii + 306pp.

22179-2 Paperbound $3.50

A HISTORY OF THE RISE AND PROGRESS OF THE ARTS OF DESIGN IN THE UNITED STATES, William Dunlap. Much the richest mine of information on early American painters, sculptors, architects, engravers, miniaturists, etc. The only source of information for scores of artists, the major primary source for many others. Unabridged reprint of rare original 1834 edition, with new introduction by James T. Flexner, and 394 new illustrations. Edited by Rita Weiss. 6⅝ x 9⅝.

21695-0, 21696-9, 21697-7 Three volumes, Paperbound $13.50

EPOCHS OF CHINESE AND JAPANESE ART, Ernest F. Fenollosa. From primitive Chinese art to the 20th century, thorough history, explanation of every important art period and form, including Japanese woodcuts; main stress on China and Japan, but Tibet, Korea also included. Still unexcelled for its detailed, rich coverage of cultural background, aesthetic elements, diffusion studies, particularly of the historical period. 2nd, 1913 edition. 242 illustrations. lii + 439pp. of text.

20364-6, 20365-4 Two volumes, Paperbound $6.00

THE GENTLE ART OF MAKING ENEMIES, James A. M. Whistler. Greatest wit of his day deflates Oscar Wilde, Ruskin, Swinburne; strikes back at inane critics, exhibitions, art journalism; aesthetics of impressionist revolution in most striking form. Highly readable classic by great painter. Reproduction of edition designed by Whistler. Introduction by Alfred Werner. xxxvi + 334pp.

21875-9 Paperbound $3.00

VISUAL ILLUSIONS: THEIR CAUSES, CHARACTERISTICS, AND APPLICATIONS, Matthew Luckiesh. Thorough description and discussion of optical illusion, geometric and perspective, particularly; size and shape distortions, illusions of color, of motion; natural illusions; use of illusion in art and magic, industry, etc. Most useful today with op art, also for classical art. Scores of effects illustrated. Introduction by William H. Ittleson. 100 illustrations. xxi + 252pp.

21530-X Paperbound $2.00

A HANDBOOK OF ANATOMY FOR ART STUDENTS, Arthur Thomson. Thorough, virtually exhaustive coverage of skeletal structure, musculature, etc. Full text, supplemented by anatomical diagrams and drawings and by photographs of undraped figures. Unique in its comparison of male and female forms, pointing out differences of contour, texture, form. 211 figures, 40 drawings, 86 photographs. xx + 459pp. 5⅜ x 8⅜.

21163-0 Paperbound $3.50

150 MASTERPIECES OF DRAWING, Selected by Anthony Toney. Full page reproductions of drawings from the early 16th to the end of the 18th century, all beautifully reproduced: Rembrandt, Michelangelo, Dürer, Fragonard, Urs, Graf, Wouwerman, many others. First-rate browsing book, model book for artists. xviii + 150pp. 8⅜ x 11¼.

21032-4 Paperbound $2.50

THE LATER WORK OF AUBREY BEARDSLEY, Aubrey Beardsley. Exotic, erotic, ironic masterpieces in full maturity: Comedy Ballet, Venus and Tannhauser, Pierrot, Lysistrata, Rape of the Lock, Savoy material, Ali Baba, Volpone, etc. This material revolutionized the art world, and is still powerful, fresh, brilliant. With *The Early Work,* all Beardsley's finest work. 174 plates, 2 in color. xiv + 176pp. 8⅛ x 11.

21817-1 Paperbound $3.00

DRAWINGS OF REMBRANDT, Rembrandt van Rijn. Complete reproduction of fabulously rare edition by Lippmann and Hofstede de Groot, completely reedited, updated, improved by Prof. Seymour Slive, Fogg Museum. Portraits, Biblical sketches, landscapes, Oriental types, nudes, episodes from classical mythology—All Rembrandt's fertile genius. Also selection of drawings by his pupils and followers. "Stunning volumes," *Saturday Review.* 550 illustrations. lxxviii + 552pp. 9⅛ x 12¼.

21485-0, 21486-9 Two volumes, Paperbound $10.00

THE DISASTERS OF WAR, Francisco Goya. One of the masterpieces of Western civilization—83 etchings that record Goya's shattering, bitter reaction to the Napoleonic war that swept through Spain after the insurrection of 1808 and to war in general. Reprint of the first edition, with three additional plates from Boston's Museum of Fine Arts. All plates facsimile size. Introduction by Philip Hofer, Fogg Museum. v + 97pp. 9⅜ x 8¼.

21872-4 Paperbound $2.00

GRAPHIC WORKS OF ODILON REDON. Largest collection of Redon's graphic works ever assembled: 172 lithographs, 28 etchings and engravings, 9 drawings. These include some of his most famous works. All the plates from *Odilon Redon: oeuvre graphique complet,* plus additional plates. New introduction and caption translations by Alfred Werner. 209 illustrations. xxvii + 209pp. 9⅛ x 12¼.

21966-8 Paperbound $4.50

DESIGN BY ACCIDENT; A BOOK OF "ACCIDENTAL EFFECTS" FOR ARTISTS AND DESIGNERS, James F. O'Brien. Create your own unique, striking, imaginative effects by "controlled accident" interaction of materials: paints and lacquers, oil and water based paints, splatter, crackling materials, shatter, similar items. Everything you do will be different; first book on this limitless art, so useful to both fine artist and commercial artist. Full instructions. 192 plates showing "accidents," 8 in color. viii + 215pp. 8⅜ x 11¼.   21942-9 Paperbound $3.50

THE BOOK OF SIGNS, Rudolf Koch. Famed German type designer draws 493 beautiful symbols: religious, mystical, alchemical, imperial, property marks, runes, etc. Remarkable fusion of traditional and modern. Good for suggestions of timelessness, smartness, modernity. Text. vi + 104pp. 6⅛ x 9¼.
                                20162-7 Paperbound $1.25

HISTORY OF INDIAN AND INDONESIAN ART, Ananda K. Coomaraswamy. An unabridged republication of one of the finest books by a great scholar in Eastern art. Rich in descriptive material, history, social backgrounds; Sunga reliefs, Rajput paintings, Gupta temples, Burmese frescoes, textiles, jewelry, sculpture, etc. 400 photos. viii + 423pp. 6⅜ x 9¾.   21436-2 Paperbound $5.00

PRIMITIVE ART, Franz Boas. America's foremost anthropologist surveys textiles, ceramics, woodcarving, basketry, metalwork, etc.; patterns, technology, creation of symbols, style origins. All areas of world, but very full on Northwest Coast Indians. More than 350 illustrations of baskets, boxes, totem poles, weapons, etc. 378 pp.
                                20025-6 Paperbound $3.00

THE GENTLEMAN AND CABINET MAKER'S DIRECTOR, Thomas Chippendale. Full reprint (third edition, 1762) of most influential furniture book of all time, by master cabinetmaker. 200 plates, illustrating chairs, sofas, mirrors, tables, cabinets, plus 24 photographs of surviving pieces. Biographical introduction by N. Bienenstock. vi + 249pp. 9⅞ x 12¾.   21601-2 Paperbound $4.00

AMERICAN ANTIQUE FURNITURE, Edgar G. Miller, Jr. The basic coverage of all American furniture before 1840. Individual chapters cover type of furniture—clocks, tables, sideboards, etc.—chronologically, with inexhaustible wealth of data. More than 2100 photographs, all identified, commented on. Essential to all early American collectors. Introduction by H. E. Keyes. vi + 1106pp. 7⅞ x 10¾.
                                21599-7, 21600-4 Two volumes, Paperbound $11.00

PENNSYLVANIA DUTCH AMERICAN FOLK ART, Henry J. Kauffman. 279 photos, 28 drawings of tulipware, Fraktur script, painted tinware, toys, flowered furniture, quilts, samplers, hex signs, house interiors, etc. Full descriptive text. Excellent for tourist, rewarding for designer, collector. Map. 146pp. 7⅞ x 10¾.
                                21205-X Paperbound $2.50

EARLY NEW ENGLAND GRAVESTONE RUBBINGS, Edmund V. Gillon, Jr. 43 photographs, 226 carefully reproduced rubbings show heavily symbolic, sometimes macabre early gravestones, up to early 19th century. Remarkable early American primitive art, occasionally strikingly beautiful; always powerful. Text. xxvi + 207pp. 8⅜ x 11¼.   21380-3 Paperbound $3.50

ALPHABETS AND ORNAMENTS, Ernst Lehner. Well-known pictorial source for decorative alphabets, script examples, cartouches, frames, decorative title pages, calligraphic initials, borders, similar material. 14th to 19th century, mostly European. Useful in almost any graphic arts designing, varied styles. 750 illustrations. 256pp. 7 x 10.   21905-4 Paperbound $4.00

PAINTING: A CREATIVE APPROACH, Norman Colquhoun. For the beginner simple guide provides an instructive approach to painting: major stumbling blocks for beginner; overcoming them, technical points; paints and pigments; oil painting; watercolor and other media and color. New section on "plastic" paints. Glossary. Formerly *Paint Your Own Pictures*. 221pp.   22000-1 Paperbound $1.75

THE ENJOYMENT AND USE OF COLOR, Walter Sargent. Explanation of the relations between colors themselves and between colors in nature and art, including hundreds of little-known facts about color values, intensities, effects of high and low illumination, complementary colors. Many practical hints for painters, references to great masters. 7 color plates, 29 illustrations. x + 274pp.
20944-X Paperbound $2.75

THE NOTEBOOKS OF LEONARDO DA VINCI, compiled and edited by Jean Paul Richter. 1566 extracts from original manuscripts reveal the full range of Leonardo's versatile genius: all his writings on painting, sculpture, architecture, anatomy, astronomy, geography, topography, physiology, mining, music, etc., in both Italian and English, with 186 plates of manuscript pages and more than 500 additional drawings. Includes studies for the Last Supper, the lost Sforza monument, and other works. Total of xlvii + 866pp. 7⅞ x 10¾.
22572-0, 22573-9 Two volumes, Paperbound $10.00

MONTGOMERY WARD CATALOGUE OF 1895. Tea gowns, yards of flannel and pillow-case lace, stereoscopes, books of gospel hymns, the New Improved Singer Sewing Machine, side saddles, milk skimmers, straight-edged razors, high-button shoes, spittoons, and on and on . . . listing some 25,000 items, practically all illustrated. Essential to the shoppers of the 1890's, it is our truest record of the spirit of the period. Unaltered reprint of Issue No. 57, Spring and Summer 1895. Introduction by Boris Emmet. Innumerable illustrations. xiii + 624pp. 8½ x 11⅝.
22377-9 Paperbound $6.95

THE CRYSTAL PALACE EXHIBITION ILLUSTRATED CATALOGUE (LONDON, 1851). One of the wonders of the modern world—the Crystal Palace Exhibition in which all the nations of the civilized world exhibited their achievements in the arts and sciences—presented in an equally important illustrated catalogue. More than 1700 items pictured with accompanying text—ceramics, textiles, cast-iron work, carpets, pianos, sleds, razors, wall-papers, billiard tables, beehives, silverware and hundreds of other artifacts—represent the focal point of Victorian culture in the Western World. Probably the largest collection of Victorian decorative art ever assembled—indispensable for antiquarians and designers. Unabridged republication of the Art-Journal Catalogue of the Great Exhibition of 1851, with all terminal essays. New introduction by John Gloag, F.S.A. xxxiv + 426pp. 9 x 12.
22503-8 Paperbound $5.00

A HISTORY OF COSTUME, Carl Köhler. Definitive history, based on surviving pieces of clothing primarily, and paintings, statues, etc. secondarily. Highly readable text, supplemented by 594 illustrations of costumes of the ancient Mediterranean peoples, Greece and Rome, the Teutonic prehistoric period; costumes of the Middle Ages, Renaissance, Baroque, 18th and 19th centuries. Clear, measured patterns are provided for many clothing articles. Approach is practical throughout. Enlarged by Emma von Sichart. 464pp.   21030-8 Paperbound $3.50

ORIENTAL RUGS, ANTIQUE AND MODERN, Walter A. Hawley. A complete and authoritative treatise on the Oriental rug—where they are made, by whom and how, designs and symbols, characteristics in detail of the six major groups, how to distinguish them and how to buy them. Detailed technical data is provided on periods, weaves, warps, wefts, textures, sides, ends and knots, although no technical background is required for an understanding. 11 color plates, 80 halftones, 4 maps. vi + 320pp. 6⅛ x 9⅛.   22366-3 Paperbound $5.00

TEN BOOKS ON ARCHITECTURE, Vitruvius. By any standards the most important book on architecture ever written. Early Roman discussion of aesthetics of building, construction methods, orders, sites, and every other aspect of architecture has inspired, instructed architecture for about 2,000 years. Stands behind Palladio, Michelangelo, Bramante, Wren, countless others. Definitive Morris H. Morgan translation. 68 illustrations. xii + 331pp.   20645-9 Paperbound $3.00

THE FOUR BOOKS OF ARCHITECTURE, Andrea Palladio. Translated into every major Western European language in the two centuries following its publication in 1570, this has been one of the most influential books in the history of architecture. Complete reprint of the 1738 Isaac Ware edition. New introduction by Adolf Placzek, Columbia Univ. 216 plates. xxii + 110pp. of text. 9½ x 12¾.   21308-0 Clothbound $12.50

STICKS AND STONES: A STUDY OF AMERICAN ARCHITECTURE AND CIVILIZATION, Lewis Mumford. One of the great classics of American cultural history. American architecture from the medieval-inspired earliest forms to the early 20th century evolution of structure and style, and reciprocal influences on environment. 21 photographic illustrations. 238pp.   20202-X Paperbound $2.00

THE AMERICAN BUILDER'S COMPANION, Asher Benjamin. The most widely used early 19th century architectural style and source book, for colonial up into Greek Revival periods. Extensive development of geometry of carpentering, construction of sashes, frames, doors, stairs; plans and elevations of domestic and other buildings. Hundreds of thousands of houses were built according to this book, now invaluable to historians, architects, restorers, etc. 1827 edition. 59 plates. 114pp. 7⅞ x 10¾.   22236-5 Paperbound $3.50

DUTCH HOUSES IN THE HUDSON VALLEY BEFORE 1776, Helen Wilkinson Reynolds. The standard survey of the Dutch colonial house and outbuildings, with constructional features, decoration, and local history associated with individual homesteads. Introduction by Franklin D. Roosevelt. Map. 150 illustrations. 469pp. 6⅝ x 9¼.   21469-9 Paperbound $5.00

CATALOGUE OF DOVER BOOKS

THE ARCHITECTURE OF COUNTRY HOUSES, Andrew J. Downing. Together with Vaux's *Villas and Cottages* this is the basic book for Hudson River Gothic architecture of the middle Victorian period. Full, sound discussions of general aspects of housing, architecture, style, decoration, furnishing, together with scores of detailed house plans, illustrations of specific buildings, accompanied by full text. Perhaps the most influential single American architectural book. 1850 edition. Introduction by J. Stewart Johnson. 321 figures, 34 architectural designs. xvi + 560pp.
22003-6 Paperbound $4.00

LOST EXAMPLES OF COLONIAL ARCHITECTURE, John Mead Howells. Full-page photographs of buildings that have disappeared or been so altered as to be denatured, including many designed by major early American architects. 245 plates. xvii + 248pp. 7⅞ x 10¾.
21143-6 Paperbound $3.50

DOMESTIC ARCHITECTURE OF THE AMERICAN COLONIES AND OF THE EARLY REPUBLIC, Fiske Kimball. Foremost architect and restorer of Williamsburg and Monticello covers nearly 200 homes between 1620-1825. Architectural details, construction, style features, special fixtures, floor plans, etc. Generally considered finest work in its area. 219 illustrations of houses, doorways, windows, capital mantels. xx + 314pp. 7⅞ x 10¾.
21743-4 Paperbound $4.00

EARLY AMERICAN ROOMS: 1650-1858, edited by Russell Hawes Kettell. Tour of 12 rooms, each representative of a different era in American history and each furnished, decorated, designed and occupied in the style of the era. 72 plans and elevations, 8-page color section, etc., show fabrics, wall papers, arrangements, etc. Full descriptive text. xvii + 200pp. of text. 8⅜ x 11¼.
21633-0 Paperbound $5.00

THE FITZWILLIAM VIRGINAL BOOK, edited by J. Fuller Maitland and W. B. Squire. Full modern printing of famous early 17th-century ms. volume of 300 works by Morley, Byrd, Bull, Gibbons, etc. For piano or other modern keyboard instrument; easy to read format. xxxvi + 938pp. 8⅜ x 11.
21068-5, 21069-3 Two volumes, Paperbound $10.00

KEYBOARD MUSIC, Johann Sebastian Bach. Bach Gesellschaft edition. A rich collection of Bach's masterpieces for the harpsichord: the six English Suites, six French Suites, the six Partitas (Clavierübung part I), the Goldberg Variations (Clavierübung part IV), the fifteen Two-Part Inventions and the fifteen Three-Part Sinfonias. Clearly reproduced on large sheets with ample margins; eminently playable. vi + 312pp. 8⅛ x 11.
22360-4 Paperbound $5.00

THE MUSIC OF BACH: AN INTRODUCTION, Charles Sanford Terry. A fine, non-technical introduction to Bach's music, both instrumental and vocal. Covers organ music, chamber music, passion music, other types. Analyzes themes, developments, innovations. x + 114pp.
21075-8 Paperbound $1.50

BEETHOVEN AND HIS NINE SYMPHONIES, Sir George Grove. Noted British musicologist provides best history, analysis, commentary on symphonies. Very thorough, rigorously accurate; necessary to both advanced student and amateur music lover. 436 musical passages. vii + 407 pp.
20334-4 Paperbound $2.75

JOHANN SEBASTIAN BACH, Philipp Spitta. One of the great classics of musicology, this definitive analysis of Bach's music (and life) has never been surpassed. Lucid, nontechnical analyses of hundreds of pieces (30 pages devoted to St. Matthew Passion, 26 to B Minor Mass). Also includes major analysis of 18th-century music, 450 musical examples. 40-page musical supplement. Total of xx + 1799pp.
(EUK) 22278-0, 22279-9 Two volumes, Clothbound $17.50

MOZART AND HIS PIANO CONCERTOS, Cuthbert Girdlestone. The only full-length study of an important area of Mozart's creativity. Provides detailed analyses of all 23 concertos, traces inspirational sources. 417 musical examples. Second edition. 509pp.
21271-8 Paperbound $3.5

THE PERFECT WAGNERITE: A COMMENTARY ON THE NIBLUNG'S RING, George Bernard Shaw. Brilliant and still relevant criticism in remarkable essays on Wagner's Ring cycle, Shaw's ideas on political and social ideology behind the plots, role of Leitmotifs, vocal requisites, etc. Prefaces. xxi + 136pp.
(USO) 21707-8 Paperbound $1.5

DON GIOVANNI, W. A. Mozart. Complete libretto, modern English translation; biographies of composer and librettist; accounts of early performances and critical reaction. Lavishly illustrated. All the material you need to understand and appreciate this great work. Dover Opera Guide and Libretto Series; translated and introduced by Ellen Bleiler. 92 illustrations. 209pp.
21134-7 Paperbound $2.0

BASIC ELECTRICITY, U. S. Bureau of Naval Personel. Originally a training course, best non-technical coverage of basic theory of electricity and its applications. Fundamental concepts, batteries, circuits, conductors and wiring techniques, AC and DC, inductance and capacitance, generators, motors, transformers, magnetic amplifier, synchros, servomechanisms, etc. Also covers blue-prints, electrical diagrams, etc. Many questions, with answers. 349 illustrations. x + 448pp. 6½ x 9¼.
20973-3 Paperbound $3.5

REPRODUCTION OF SOUND, Edgar Villchur. Thorough coverage for laymen of high fidelity systems, reproducing systems in general, needles, amplifiers, preamps, loudspeakers, feedback, explaining physical background. "A rare talent for making technicalities vividly comprehensible," R. Darrell, *High Fidelity*. 69 figures. iv + 92pp.
21515-6 Paperbound $1.25

HEAR ME TALKIN' TO YA: THE STORY OF JAZZ AS TOLD BY THE MEN WHO MADE IT, Nat Shapiro and Nat Hentoff. Louis Armstrong, Fats Waller, Jo Jones, Clarence Williams, Billy Holiday, Duke Ellington, Jelly Roll Morton and dozens of other jazz greats tell how it was in Chicago's South Side, New Orleans, depression Harlem and the modern West Coast as jazz was born and grew. xvi + 429pp.
21726-4 Paperbound $3.0

FABLES OF AESOP, translated by Sir Roger L'Estrange. A reproduction of the very rare 1931 Paris edition; a selection of the most interesting fables, together with 50 imaginative drawings by Alexander Calder. v + 128pp. 6½x9¼.
21780-9 Paperbound $1.5

CATALOGUE OF DOVER BOOKS

AGAINST THE GRAIN (A REBOURS), Joris K. Huysmans. Filled with weird images, evidences of a bizarre imagination, exotic experiments with hallucinatory drugs, rich tastes and smells and the diversions of its sybarite hero Duc Jean des Esseintes, this classic novel pushed 19th-century literary decadence to its limits. Full unabridged edition. Do not confuse this with abridged editions generally sold. Introduction by Havelock Ellis. xlix + 206pp.   22190-3 Paperbound $2.00

VARIORUM SHAKESPEARE: HAMLET. Edited by Horace H. Furness; a landmark of American scholarship. Exhaustive footnotes and appendices treat all doubtful words and phrases, as well as suggested critical emendations throughout the play's history. First volume contains editor's own text, collated with all Quartos and Folios. Second volume contains full first Quarto, translations of Shakespeare's sources (Belleforest, and Saxo Grammaticus), Der Bestrafte Brudermord, and many essays on critical and historical points of interest by major authorities of past and present. Includes details of staging and costuming over the years. By far the best edition available for serious students of Shakespeare. Total of xx + 905pp.
21004-9, 21005-7, 2 volumes, Paperbound $7.00

A LIFE OF WILLIAM SHAKESPEARE, Sir Sidney Lee. This is the standard life of Shakespeare, summarizing everything known about Shakespeare and his plays. Incredibly rich in material, broad in coverage, clear and judicious, it has served thousands as the best introduction to Shakespeare. 1931 edition. 9 plates. xxix + 792pp.   (USO) 21967-4 Paperbound $3.75

MASTERS OF THE DRAMA, John Gassner. Most comprehensive history of the drama in print, covering every tradition from Greeks to modern Europe and America, including India, Far East, etc. Covers more than 800 dramatists, 2000 plays, with biographical material, plot summaries, theatre history, criticism, etc. "Best of its kind in English," New Republic. 77 illustrations. xxii + 890pp.
20100-7 Clothbound $8.50

THE EVOLUTION OF THE ENGLISH LANGUAGE, George McKnight. The growth of English, from the 14th century to the present. Unusual, non-technical account presents basic information in very interesting form: sound shifts, change in grammar and syntax, vocabulary growth, similar topics. Abundantly illustrated with quotations. Formerly Modern English in the Making. xii + 590pp.
21932-1 Paperbound $3.50

AN ETYMOLOGICAL DICTIONARY OF MODERN ENGLISH, Ernest Weekley. Fullest, richest work of its sort, by foremost British lexicographer. Detailed word histories, including many colloquial and archaic words; extensive quotations. Do not confuse this with the Concise Etymological Dictionary, which is much abridged. Total of xxvii + 830pp. 6½ x 9¼.
21873-2, 21874-0 Two volumes, Paperbound $7.90

FLATLAND: A ROMANCE OF MANY DIMENSIONS, E. A. Abbott. Classic of science-fiction explores ramifications of life in a two-dimensional world, and what happens when a three-dimensional being intrudes. Amusing reading, but also useful as introduction to thought about hyperspace. Introduction by Banesh Hoffmann. 16 illustrations. xx + 103pp.   20001-9 Paperbound $1.00

POEMS OF ANNE BRADSTREET, edited with an introduction by Robert Hutchinson. A new selection of poems by America's first poet and perhaps the first significant woman poet in the English language. 48 poems display her development in works of considerable variety—love poems, domestic poems, religious meditations, formal elegies, "quaternions," etc. Notes, bibliography. viii + 222pp.

22160-1 Paperbound $2.50

THREE GOTHIC NOVELS: THE CASTLE OF OTRANTO BY HORACE WALPOLE; VATHEK BY WILLIAM BECKFORD; THE VAMPYRE BY JOHN POLIDORI, WITH FRAGMENT OF A NOVEL BY LORD BYRON, edited by E. F. Bleiler. The first Gothic novel, by Walpole; the finest Oriental tale in English, by Beckford; powerful Romantic supernatural story in versions by Polidori and Byron. All extremely important in history of literature; all still exciting, packed with supernatural thrills, ghosts, haunted castles, magic, etc. xl + 291pp.

21232-7 Paperbound $2.50

THE BEST TALES OF HOFFMANN, E. T. A. Hoffmann. 10 of Hoffmann's most important stories, in modern re-editings of standard translations: Nutcracker and the King of Mice, Signor Formica, Automata, The Sandman, Rath Krespel, The Golden Flowerpot, Master Martin the Cooper, The Mines of Falun, The King's Betrothed, A New Year's Eve Adventure. 7 illustrations by Hoffmann. Edited by E. F. Bleiler. xxxix + 419pp.

21793-0 Paperbound $3.00

GHOST AND HORROR STORIES OF AMBROSE BIERCE, Ambrose Bierce. 23 strikingly modern stories of the horrors latent in the human mind: The Eyes of the Panther, The Damned Thing, An Occurrence at Owl Creek Bridge, An Inhabitant of Carcosa, etc., plus the dream-essay, Visions of the Night. Edited by E. F. Bleiler. xxii + 199pp.

20767-6 Paperbound $1.50

BEST GHOST STORIES OF J. S. LEFANU, J. Sheridan LeFanu. Finest stories by Victorian master often considered greatest supernatural writer of all. Carmilla, Green Tea, The Haunted Baronet, The Familiar, and 12 others. Most never before available in the U. S. A. Edited by E. F. Bleiler. 8 illustrations from Victorian publications. xvii + 467pp.

20415-4 Paperbound $3.00

MATHEMATICAL FOUNDATIONS OF INFORMATION THEORY, A. I. Khinchin. Comprehensive introduction to work of Shannon, McMillan, Feinstein and Khinchin, placing these investigations on a rigorous mathematical basis. Covers entropy concept in probability theory, uniqueness theorem, Shannon's inequality, ergodic sources, the E property, martingale concept, noise, Feinstein's fundamental lemma, Shanon's first and second theorems. Translated by R. A. Silverman and M. D. Friedman. iii + 120pp.

60434-9 Paperbound $2.00

SEVEN SCIENCE FICTION NOVELS, H. G. Wells. The standard collection of the great novels. Complete, unabridged. *First Men in the Moon, Island of Dr. Moreau, War of the Worlds, Food of the Gods, Invisible Man, Time Machine, In the Days of the Comet.* Not only science fiction fans, but every educated person owes it to himself to read these novels. 1015pp.

(USO) 20264-X Clothbound $6.00

ᴀsᴛ ᴀɴᴅ Fɪʀsᴛ Mᴇɴ ᴀɴᴅ Sᴛᴀʀ Mᴀᴋᴇʀ, Tᴡᴏ Sᴄɪᴇɴᴄᴇ Fɪᴄᴛɪᴏɴ Nᴏᴠᴇʟs, Olaf
ᴛapledon. Greatest future histories in science fiction. In the first, human intelli-
ᴇnce is the "hero," through strange paths of evolution, interplanetary invasions,
ᴀcredible technologies, near extinctions and reemergences. Star Maker describes the
ᴜest of a band of star rovers for intelligence itself, through time and space: weird
ᴀhuman civilizations, crustacean minds, symbiotic worlds, etc. Complete, un-
bridged. v + 438pp. (USO) 21962-3 Paperbound $2.50

ʜʀᴇᴇ Pʀᴏᴘʜᴇᴛɪᴄ Nᴏᴠᴇʟs, H. G. Wᴇʟʟs. Stages of a consistently planned future
ᴏr mankind. *When the Sleeper Wakes,* and *A Story of the Days to Come,* anticipate
ᴊave New World and *1984,* in the 21st Century; *The Time Machine,* only com-
ᴌete version in print, shows farther future and the end of mankind. All show
Jells's greatest gifts as storyteller and novelist. Edited by E. F. Bleiler. x
- 335pp. (USO) 20605-X Paperbound $2.50

ʜᴇ Dᴇᴠɪʟ's Dɪᴄᴛɪᴏɴᴀʀʏ, Ambrose Bierce. America's own Oscar Wilde—
ᴍbrose Bierce—offers his barbed iconoclastic wisdom in over 1,000 definitions
ᴀiled by H. L. Mencken as "some of the most gorgeous witticisms in the English
ᴀguage." 145pp. 20487-1 Paperbound $1.25

ᴀx ᴀɴᴅ Mᴏʀɪᴛᴢ, Wilhelm Busch. Great children's classic, father of comic
ᴛrip, of two bad boys, Max and Moritz. Also Ker and Plunk (Plisch und Plumm),
ᴀt and Mouse, Deceitful Henry, Ice-Peter, The Boy and the Pipe, and five other
ᴇces. Original German, with English translation. Edited by H. Arthur Klein;
ᴀnslations by various hands and H. Arthur Klein. vi + 216pp.
20181-3 Paperbound $2.00

ᴀGs ɪs PɪGs ᴀɴᴅ Oᴛʜᴇʀ Fᴀᴠᴏʀɪᴛᴇs, Ellis Parker Butler. The title story is one
ᴼ the best humor short stories, as Mike Flannery obfuscates biology and English.
ᴌso included, That Pup of Murchison's, The Great American Pie Company, and
ᴇrkins of Portland. 14 illustrations. v + 109pp. 21532-6 Paperbound $1.25

ʜᴇ Pᴇᴛᴇʀᴋɪɴ Pᴀᴘᴇʀs, Lucretia P. Hale. It takes genius to be as stupidly mad as
ᴇ Peterkins, as they decide to become wise, celebrate the "Fourth," keep a cow,
ᴀd otherwise strain the resources of the Lady from Philadelphia. Basic book of
ᴍerican humor. 153 illustrations. 219pp. 20794-3 Paperbound $1.50

ᴇʀʀᴀᴜʟᴛ's Fᴀɪʀʏ Tᴀʟᴇs, translated by A. E. Johnson and S. R. Littlewood, with
ᴀ full-page illustrations by Gustave Doré. All the original Perrault stories—
ᴀnderella, Sleeping Beauty, Bluebeard, Little Red Riding Hood, Puss in Boots, Tom
ʜumb, etc.—with their witty verse morals and the magnificent illustrations of
ᴏré. One of the five or six great books of European fairy tales. viii + 117pp.
ᴶ₈ x 11. 22311-6 Paperbound $2.00

ʟᴅ HᴜɴGᴀʀɪᴀɴ Fᴀɪʀʏ Tᴀʟᴇs, Baroness Orczy. Favorites translated and adapted
ᴼ author of the *Scarlet Pimpernel.* Eight fairy tales include "The Suitors of Princess
ᴊe-Fly," "The Twin Hunchbacks," "Mr. Cuttlefish's Love Story," and "The
ᴀchanted Cat." This little volume of magic and adventure will captivate children
ᴀit has for generations. 90 drawings by Montagu Barstow. 96pp.
22293-4 Paperbound $1.95

THE RED FAIRY BOOK, Andrew Lang. Lang's color fairy books have long bee children's favorites. This volume includes Rapunzel, Jack and the Bean-stalk an 35 other stories, familiar and unfamiliar. 4 plates, 93 illustrations x + 367pp.
21673-X Paperbound $2.5

THE BLUE FAIRY BOOK, Andrew Lang. Lang's tales come from all countries and a times. Here are 37 tales from Grimm, the Arabian Nights, Greek Mythology, an other fascinating sources. 8 plates, 130 illustrations. xi + 390pp.
21437-0 Paperbound $2.5

HOUSEHOLD STORIES BY THE BROTHERS GRIMM. Classic English-language editio of the well-known tales — Rumpelstiltskin, Snow White, Hansel and Gretel, Th Twelve Brothers, Faithful John, Rapunzel, Tom Thumb (52 stories in all). Trans lated into simple, straightforward English by Lucy Crane. Ornamented with head pieces, vignettes, elaborate decorative initials and a dozen full-page illustrations b Walter Crane. x + 269pp.
21080-4 Paperbound $2.0

THE MERRY ADVENTURES OF ROBIN HOOD, Howard Pyle. The finest modern ver sions of the traditional ballads and tales about the great English outlaw. Howar Pyle's complete prose version, with every word, every illustration of the first editio Do not confuse this facsimile of the original (1883) with modern editions tha change text or illustrations. 23 plates plus many page decorations. xxii + 296pp
22043-5 Paperbound $2.5

THE STORY OF KING ARTHUR AND HIS KNIGHTS, Howard Pyle. The finest chil dren's version of the life of King Arthur; brilliantly retold by Pyle, with 48 of hi most imaginative illustrations. xviii + 313pp. 6⅛ x 9¼.
21445-1 Paperbound $2.5

THE WONDERFUL WIZARD OF OZ, L. Frank Baum. America's finest children' book in facsimile of first edition with all Denslow illustrations in full color. Th edition a child should have. Introduction by Martin Gardner. 23 color plates scores of drawings. iv + 267pp.
20691-2 Paperbound $2.5

THE MARVELOUS LAND OF OZ, L. Frank Baum. The second Oz book, every bit a imaginative as the Wizard. The hero is a boy named Tip, but the Scarecrow and th Tin Woodman are back, as is the Oz magic. 16 color plates, 120 drawings by Joh R. Neill. 287pp.
20692-0 Paperbound $2.5

THE MAGICAL MONARCH OF MO, L. Frank Baum. Remarkable adventures in a lan even stranger than Oz. The best of Baum's books not in the Oz series. 15 colo plates and dozens of drawings by Frank Verbeck. xviii + 237pp.
21892-9 Paperbound $2.2

THE BAD CHILD'S BOOK OF BEASTS, MORE BEASTS FOR WORSE CHILDREN, MORAL ALPHABET, Hilaire Belloc. Three complete humor classics in one volume Be kind to the frog, and do not call him names . . . and 28 other whimsical animal Familiar favorites and some not so well known. Illustrated by Basil Blackwel. 156pp.
(USO) 20749-8 Paperbound $1.5

EAST O' THE SUN AND WEST O' THE MOON, George W. Dasent. Considered the best of all translations of these Norwegian folk tales, this collection has been enjoyed by generations of children (and folklorists too). Includes True and Untrue, Why the Sea is Salt, East O' the Sun and West O' the Moon, Why the Bear is Stumpy-Tailed, Boots and the Troll, The Cock and the Hen, Rich Peter the Pedlar, and 52 more. The only edition with all 59 tales. 77 illustrations by Erik Werenskiold and Theodor Kittelsen. xv + 418pp. 22521-6 Paperbound $3.50

GOOPS AND HOW TO BE THEM, Gelett Burgess. Classic of tongue-in-cheek humor, masquerading as etiquette book. 87 verses, twice as many cartoons, show mischievous Goops as they demonstrate to children virtues of table manners, neatness, courtesy, etc. Favorite for generations. viii + 88pp. 6½ x 9¼. 22233-0 Paperbound $1.25

ALICE'S ADVENTURES UNDER GROUND, Lewis Carroll. The first version, quite different from the final *Alice in Wonderland,* printed out by Carroll himself with his own illustrations. Complete facsimile of the "million dollar" manuscript Carroll gave to Alice Liddell in 1864. Introduction by Martin Gardner. viii + 96pp. Title and dedication pages in color. 21482-6 Paperbound $1.25

THE BROWNIES, THEIR BOOK, Palmer Cox. Small as mice, cunning as foxes, exuberant and full of mischief, the Brownies go to the zoo, toy shop, seashore, circus, etc., in 24 verse adventures and 266 illustrations. Long a favorite, since their first appearance in St. Nicholas Magazine. xi + 144pp. 6⅝ x 9¼. 21265-3 Paperbound $1.75

SONGS OF CHILDHOOD, Walter De La Mare. Published (under the pseudonym Walter Ramal) when De La Mare was only 29, this charming collection has long been a favorite children's book. A facsimile of the first edition in paper, the 47 poems capture the simplicity of the nursery rhyme and the ballad, including such lyrics as I Met Eve, Tartary, The Silver Penny. vii + 106pp. (USO) 21972-0 Paperbound $1.25

THE COMPLETE NONSENSE OF EDWARD LEAR, Edward Lear. The finest 19th-century humorist-cartoonist in full: all nonsense limericks, zany alphabets, Owl and Pussycat, songs, nonsense botany, and more than 500 illustrations by Lear himself. Edited by Holbrook Jackson. xxix + 287pp. (USO) 20167-8 Paperbound $2.00

BILLY WHISKERS: THE AUTOBIOGRAPHY OF A GOAT, Frances Trego Montgomery. A favorite of children since the early 20th century, here are the escapades of that rambunctious, irresistible and mischievous goat—Billy Whiskers. Much in the spirit of *Peck's Bad Boy,* this is a book that children never tire of reading or hearing. All the original familiar illustrations by W. H. Fry are included: 6 color plates, 50 black and white drawings. 159pp. 22345-0 Paperbound $2.00

MOTHER GOOSE MELODIES. Faithful republication of the fabulously rare Munroe and Francis "copyright 1833" Boston edition—the most important Mother Goose collection, usually referred to as the "original." Familiar rhymes plus many rare ones, with wonderful old woodcut illustrations. Edited by E. F. Bleiler. 128pp. 4½ x 6⅜. 22577-1 Paperbound $1.00

TWO LITTLE SAVAGES; BEING THE ADVENTURES OF TWO BOYS WHO LIVED A‹ INDIANS AND WHAT THEY LEARNED, Ernest Thompson Seton. Great classic c nature and boyhood provides a vast range of woodlore in most palatable form, genuinely entertaining story. Two farm boys build a teepee in woods and live in ‹ for a month, working out Indian solutions to living problems, star lore, birds an‹ animals, plants, etc. 293 illustrations. vii + 286pp.

20985-7 Paperbound $2.5‹

PETER PIPER'S PRACTICAL PRINCIPLES OF PLAIN & PERFECT PRONUNCIATION‹ Alliterative jingles and tongue-twisters of surprising charm, that made their fir‹ appearance in America about 1830. Republished in full with the spirited woodc‹ illustrations from this earliest American edition. 32pp. 4½ x 6⅜.

22560-7 Paperbound $1.C‹

SCIENCE EXPERIMENTS AND AMUSEMENTS FOR CHILDREN, Charles Vivian. 73 ea‹ experiments, requiring only materials found at home or easily available, such ‹ candles, coins, steel wool, etc.; illustrate basic phenomena like vacuum, simp‹ chemical reaction, etc. All safe. Modern, well-planned. Formerly *Science Gam‹ for Children.* 102 photos, numerous drawings. 96pp. 6⅛ x 9¼.

21856-2 Paperbound $1.2‹

AN INTRODUCTION TO CHESS MOVES AND TACTICS SIMPLY EXPLAINED, Leona‹ Barden. Informal intermediate introduction, quite strong in explaining reasons f‹ moves. Covers basic material, tactics, important openings, traps, positional play‹ middle game, end game. Attempts to isolate patterns and recurrent configuration‹ Formerly *Chess.* 58 figures. 102pp. (USO) 21210-6 Paperbound $1.‹

LASKER'S MANUAL OF CHESS, Dr. Emanuel Lasker. Lasker was not only one of t‹ five great World Champions, he was also one of the ablest expositors, theorists, a‹ analysts. In many ways, his Manual, permeated with his philosophy of battle, fill‹ with keen insights, is one of the greatest works ever written on chess. Filled w‹ analyzed games by the great players. A single-volume library that will profit alm‹ any chess player, beginner or master. 308 diagrams. xli x 349pp.

20640-8 Paperbound $2.‹

THE MASTER BOOK OF MATHEMATICAL RECREATIONS, Fred Schuh. In opinion‹ many the finest work ever prepared on mathematical puzzles, stunts, recreation‹ exhaustively thorough explanations of mathematics involved, analysis of effec‹ citation of puzzles and games. Mathematics involved is elementary. Translated‹ F. Göbel. 194 figures. xxiv + 430pp. 22134-2 Paperbound $3‹

MATHEMATICS, MAGIC AND MYSTERY, Martin Gardner. Puzzle editor for Scienti‹ American explains mathematics behind various mystifying tricks: card tricks, sta‹ "mind reading," coin and match tricks, counting out games, geometric dissectio‹ etc. Probability sets, theory of numbers clearly explained. Also provides more th‹ 400 tricks, guaranteed to work, that you can do. 135 illustrations. xii + 176pp.

20335-2 Paperbound $1.‹

MATHEMATICAL PUZZLES FOR BEGINNERS AND ENTHUSIASTS, Geoffrey Mott-Smith. 189 puzzles from easy to difficult—involving arithmetic, logic, algebra, properties of digits, probability, etc.—for enjoyment and mental stimulus. Explanation of mathematical principles behind the puzzles. 135 illustrations. viii + 248pp.
20198-8 Paperbound $1.75

PAPER FOLDING FOR BEGINNERS, William D. Murray and Francis J. Rigney. Easiest book on the market, clearest instructions on making interesting, beautiful origami. Sail boats, cups, roosters, frogs that move legs, bonbon boxes, standing birds, etc. 40 projects; more than 275 diagrams and photographs. 94pp.
20713-7 Paperbound $1.00

TRICKS AND GAMES ON THE POOL TABLE, Fred Herrmann. 79 tricks and games—some solitaires, some for two or more players, some competitive games—to entertain you between formal games. Mystifying shots and throws, unusual caroms, tricks involving such props as cork, coins, a hat, etc. Formerly *Fun on the Pool Table*. 77 figures. 95pp.
21814-7 Paperbound $1.00

HAND SHADOWS TO BE THROWN UPON THE WALL: A SERIES OF NOVEL AND AMUSING FIGURES FORMED BY THE HAND, Henry Bursill. Delightful picturebook from great-grandfather's day shows how to make 18 different hand shadows: a bird that flies, duck that quacks, dog that wags his tail, camel, goose, deer, boy, turtle, etc. Only book of its sort. vi + 33pp. 6½ x 9¼. 21779-5 Paperbound $1.00

WHITTLING AND WOODCARVING, E. J. Tangerman. 18th printing of best book on market. "If you can cut a potato you can carve" toys and puzzles, chains, chessmen, caricatures, masks, frames, woodcut blocks, surface patterns, much more. Information on tools, woods, techniques. Also goes into serious wood sculpture from Middle Ages to present, East and West. 464 photos, figures. x + 293pp.
20965-2 Paperbound $2.00

HISTORY OF PHILOSOPHY, Julián Marias. Possibly the clearest, most easily followed, best planned, most useful one-volume history of philosophy on the market; neither skimpy nor overfull. Full details on system of every major philosopher and dozens of less important thinkers from pre-Socratics up to Existentialism and later. Strong on many European figures usually omitted. Has gone through dozens of editions in Europe. 1966 edition, translated by Stanley Appelbaum and Clarence Strowbridge. xiii + 505pp.
21739-6 Paperbound $3.50

YOGA: A SCIENTIFIC EVALUATION, Kovoor T. Behanan. Scientific but non-technical study of physiological results of yoga exercises; done under auspices of Yale U. Relations to Indian thought, to psychoanalysis, etc. 16 photos. xxiii + 270pp.
20505-3 Paperbound $2.50

Prices subject to change without notice.

Available at your book dealer or write for free catalogue to Dept. GI, Dover Publications, Inc., 180 Varick St., N. Y., N. Y. 10014. Dover publishes more than 150 books each year on science, elementary and advanced mathematics, biology, music, art, literary history, social sciences and other areas.